Massih Zekavat

# Leveraging Satire for Environmental Advocacy

## Creative Arts in the Chthulucene

palgrave
macmillan

Massih Zekavat
Independent Researcher
Groningen, Netherlands

ISBN 978-3-032-04837-0 ISBN 978-3-032-04838-7 (eBook)
https://doi.org/10.1007/978-3-032-04838-7



This Palgrave Macmillan imprint is published by the registered company Springer Nature Switzerland AG.
The registered company address is: Gewerbestrasse 11, 6330 Cham, Switzerland

*For*
*Saba*

# Statement

I stand in solidarity with Palestine.

# Acknowledgment

I am grateful to Dr. Roxanne Douglas for generously taking the time to read the draft of parts of this book and offering valuable and constructive feedback. And I will remain forever grateful to my mother, Monireh Dayhimnia.

I benefited from the use of generative AI tools for language refinement, paraphrasing, copyediting, and compiling the list of references in writing this book. All intellectual contributions, critical analyses, and original arguments remain my own. Any errors or oversights are my sole responsibility.

*Competing Interests* The author has no competing interests to declare that are relevant to the content of this manuscript.

# Contents

# About the Author

**Massih Zekavat** is an independent researcher. He is author of Satire, Humor and the Construction of Identities (John Benjamins) and co-author of Satire, Humor, and Environmental Crises (Routledge).

# List of Figures

CHAPTER 1

# The Drunkard's Quest

Late at pitch-black night, a constable came across a drunk man on the street. He was clearly searching for something under the lamppost. The constable asked him, "What are you looking for?" "My keys," he responded. The constable joined him in searching to help him find his keys. After a while when they could not find anything, the officer asked the drunkard, "Are you sure you lost your keys here?" He responded, "No, actually, I lost them inside my house." "Why are you looking for them here on the street, then?" "Because I don't have a lamp at home, and here's where the light is."

This is paraphrased from a joke in Ubayd Zakani's works, a renowned fourteenth-century Persian satirist. The joke has lent itself to describing an observational bias in psychology, namely streetlight bias or the drunkard's search principle according to which we tend to engage in activities that do not require a lot of investment, regardless of their sufficiency. It can partly explain why we favor undemanding actions in addressing climate change, although we are fully aware that they are not nearly sufficient for tackling the scope and urgency of the looming crisis. Humor sheds light on intention-action and attitude-behavior gaps in this case. It can be employed in other ways like information provision, community building, or relatability in addressing the climate crisis, as well (Zekavat & Scheel, 2023).



M. Zekavat, *Leveraging Satire for Environmental Advocacy*,
https://doi.org/10.1007/978-3-032-04838-7_1

When faced with danger, animals typically exhibit one of four responses: fight, flight, freeze, or fawn. The fight response involves confronting a threat head-on, probably with aggression. Flight is characterized by the urge to escape and avoid the danger entirely. Freeze manifests as an inability to act or move in the face of a threat, while fawn involves attempting to appease the threat to prevent conflict. In the context of climate change and ecological mutations, it seems we have collectively chosen the freeze response. Despite the urgency and widespread awareness of the issue, as scientists, corporations, and politicians are well-informed and time is fleeting, significant action remains elusive. As Bruno Latour poignantly notes, "One can of course blame the inertia of habits, the fear of novelty, the heady benefits of consumerism, the iron cage of capitalism; one can point to the influence of the lobbies that work actively on disinformation; or one can take into account the work of psychosociologists on the fear that paralyzes instead of provoking a reaction" (Latour, 2017, p. 191). The basic premise of this book is how fear paralyzes instead of provoking a timely and proportionate reaction to the climate urgency. It can cause ακρασία (akrasia) or absence of self-mastery. In philosophical discourse, akrasia refers to a weakness of will or incontinence—the tendency to act in opposition to one's own deliberate judgment regarding the most rational or optimal course of action.

The fear surrounding the disconcerting scope and magnitude of ecological mutations contributes to our inaction, yet it is not the only explanation. While it is true that individuals may feel overwhelmed and powerless in their encounter with nature's formidable forces, another critical factor is material dispossession. As Latour posits in *Down to Earth* (Latour, 2018), climate *quietism* can be attributed, at least in part, to "the ordeal of finding oneself *deprived of land*." Ecological mutations and dispossession are intertwined with imperial and colonial practices that have historically inflicted violence upon not only indigenous populations but also other species and their ecological habitats.

In *Facing Gaia* (Latour, 2017), Latour offers a slightly different perspective on the roots of quietism. Contrary to the emphasis in psychology and communication science, he explicitly dismisses the idea that the scope of ecological mutations or their distance from our everyday concerns is the primary issue, stating, "Don't tell me that it's the scope of the threat or the distance from our daily preoccupations that makes the difference" (Latour, 2017, p. 191); rather, the issue lies in our departure from true materialism. He asserts that we "decided to remain insensitive to the reactions of beings of a certain type—those who are connected, broadly speaking, to the strange figure of matter. In other words, what we have to understand is why we are not true materialists" (Latour, 2017, p. 191). This insensitivity is the result of a "virile decision" to disregard potential dangers. Consequently, there is a tendency toward disinhibition regarding future actions, coupled with inhibition in confronting the retroactive consequences of those actions. Latour characterizes this dichotomy as "virility on one side, impotence on the other" (Latour, 2017, p. 192). I will return to this later to show how the disinhibition inherent in the futurity of "hope" can inhibit timely and proportionate climate action.

Latour attributes disinhibition in the face of ecological crises to a quasi-religious certainty associated with scientific understanding. The curious relationship between religion and science has contributed to the rise of apocalyptic narratives surrounding climate change, which Latour prefers to describe as "ecological mutations." He suggests that the concept of an end time, coupled with the notion of absolute truth, has been adapted into an unlikely form in the concept of progress. Certain groups of people now believe with absolute certainty that they have reached an end of time, transitioning into a new world and experiencing a complete break from the past. To these individuals, nothing significant can occur anymore, as they perceive themselves to be living within the "end of history." Consequently, warnings of an impending apocalypse hold little sway over them, as they believe they have already transcended to a different realm well beyond the apocalypse. They consider themselves to be thoroughly modernized, with their only trajectory being forward, never backward (Latour, 2017, pp. 195–196). I will further explore this inefficacy of apocalyptic narratives and their visceral and cognitive dimensions in this book, proposing satire as an alternative narrative strategy for communicating exponentially escalating anthropogenic ecological mutations.

Latour's critique of modernity and progress narratives delves deeper into the paradoxical adoption of end-time concepts by ostensibly secular societies. He observes that these societies, identifying as non-religious and secular, have ironically embraced the profound essence of the counter-religion that preceded them, specifically the notion that one can exist within an end time. However, they have inverted this idea, asserting with certainty that the end time has indeed arrived. In this transformation, essential elements such as doubt, uncertainty, and the awe-inspiring fear of time's potential cessation—and the impossibility of achieving anything without the flow of time—have been lost. According to Latour, the misunderstanding hinges on the nuanced interpretation of the term "definitive" (Latour, 2017, p. 196).

Latour further contends that there is no basis for predicting the world's end; it can only be preached or wished for. The concept of "end" encompasses telos, achievement, finitude, and revelation, yet these occur within and through the passage of time. Time itself gains new significance, as it is the medium through which final achievements—which are never truly final—are realized. Paradoxically, what is everlasting endures only through what is transient. This tension is another drawback of apocalyptic narratives that does not necessarily recur in satiric alternatives. To truly grasp this complex situation, one must not attempt to escape time. Doing so risks opposing the ongoing passage of time with the notion of time needing to end to reach permanence. This thinking characterizes millenarians or, in an unusual reversal, those who claim that the waiting period is over and history has concluded or is about to conclude. Interpreting "the time of the end" as "the end of times" leads to a profound transformation and a compelling temptation to embrace eternity, abandoning the finite and mortal experience of time (Latour, 2017, pp. 196–197).

This oversight leads to a fundamental misunderstanding of apocalyptic messages. As Latour notes, "modernization retains all the apocalyptic features but deprives itself of the uncertainty that was required to keep science, politics, and religion from getting mixed together" (Latour, 2017, p. 199). Living in an era characterized by increasing anthropogenic ecological mutations necessitates a constant awareness of existing at the brink of an ending. It is not about the futurity of a looming apocalypse that might befall if 'we' do not change course; rather, it is about living in the time of the end constantly unfolding in real time. Latour suggests that those who criticize ecological discourse for being overly apocalyptic are, in fact, the ones who have enforced a violent conclusion on other civilizations by refusing to acknowledge the continuous presence of a time of the end (Latour, 2017, p. 206).

Latour further argues that the lack of widespread interest in ecological issues, despite their immense scale, urgency, and persistence, can be traced to their (counter-)religious origins. He explains that telling Westerners—or those who have recently, and often violently, adopted Western ideals—that their time has come and that their world has ended, necessitating a change in lifestyle, results in bewilderment. For these individuals, the apocalypse has ostensibly already occurred. They believe they have transitioned to another realm, where the world beyond has been realized, particularly for the affluent. In their view, they have already surpassed the threshold that marks the end of historical progression (Latour, 2017, p. 206).

As a result, a fundamental disbelief persists despite widespread awareness of the climate urgency among individuals, corporations, and policymakers. As Latour observes, "They know, they hear, but deep down they do not believe it. Here is where we have to seek the fundamental source of climate skepticism, I believe" (Latour, 2017, p. 206). This skepticism is not rooted in doubts about the reliability of scientific knowledge, but rather in a deeper existential uncertainty. So skeptics often dismiss those advocating for a total and radical change in lifestyle as irrational or unworthy of serious consideration.

Adam Douglas Thompson's cartoon in *The New Yorker* (Thompson, 2025) captures the quintessence of denialism and quietism at the end of times, while taking a jab at apocalyptic narratives. It depicts a post-apocalyptic scene with burning cars, buildings, and trees, skulls and bones scattered. Two figures stand out in the center: one sits cross-legged with a hand-under-chin pose; the other is engrossed in a book, saying, "Lately, I've been getting really into pre-apocalyptic fiction." This cartoon rebukes the illusion of living in the end of times and highlights the tendency to deny ecological urgency. Despite the unfolding apocalypse, the figures' nonchalance is symptomatic of socio-political quietism rooted in losing sight of the time of the end. The cartoon critiques our collective denial and inaction, opting for escapism over action in the face of disaster.

Latour identifies material engagement as the missing link in addressing this issue. "What doesn't manage to get through to people bombarded by bad news about the ecological mutation is the activity, the autonomy, the sensitivity to our actions, of the materials that make up the critical zones in which we all reside. These people seem incapable of responding to the agency of these materials" (Latour, 2017, p. 207). We fail to recognize the activity, autonomy, and sensitivity of vibrant matter to human existence and actions, so the challenge lies in conveying the dynamic and responsive nature of the materials that constitute the critical zones we inhabit as holobionts.

## A Counter-Religious Conversion

Environmental psychology is an example of what Latour calls a counter-religion. It offers a diverse array of theories and models to understand environmental behavior, often characterized by two primary trends: one emphasizing rational behavior and the other focusing on affective and symbolic aspects.[1] One model is the Theory of the Meaning of Material Possessions (Gatersleben & Steg, 2013) which posits that environmental behavior is influenced by the instrumental, affective, and symbolic aspects of material products. The instrumental function refers to the practical use of a product, while its symbolic functions might include status or gender roles associated with its possession. The affective aspect pertains to the emotions—such as power or arousal—elicited by owning and using the product. These aspects are interconnected and can mutually influence environmental behavior.

In contrast, the Theory of Planned Behavior (TPB), proposed by Ajzen (1985, 1991), highlights the role of rational decision-making in environmental behavior. According to TPB, behavior is determined by intention, which is influenced by attitudes, subjective norms, and perceived behavioral control. TPB suggests that individuals' vested interests primarily drive their environmental actions.

Shifting focus from rational self-interest, the Norm-Activation Model (NAM) and the Value-Belief-Norm Theory (VBN) emphasize the role of personal norms and values. NAM, based on Schwartz's value theory (Schwartz, 1992, 1994), suggests that pro-environmental behavior is predicted by the activation of personal norms through problem awareness, responsibility ascription, outcome efficacy, and self-efficacy. VBN extends NAM by incorporating values and ecological worldviews. It posits that altruistic and biospheric values promote pro-environmental behavior, while egoistic values do not. These values shape beliefs and personal norms that lead to pro-environmental behaviors across various domains, such as activism and private sphere actions.

The Theory of Planned Behavior is particularly effective for predicting high-cost behaviors, while NAM and VBN are better suited for low-cost behaviors. To create a more comprehensive understanding, researchers have integrated elements from these theories into unified models. For instance, Goal Framing Theory combines insights from NAM and TPB, suggesting that behavior is guided by hedonic, gain, and normative goals.

[1] Abrahamse (2019), Brick and Lewis (2016), Corral-Verdugo et al. (2012), de Groot and Thøgersen (2013), de Leeuw et al. (2015), Gifford and Sussman (2012), Kals and Müller (2012), Kormos and Gifford (2014), Lindenberg (2018), Lindenberg and Steg (2007), Pfattheicher et al. (2016), Schultz and Kaiser (2012), Schultz et al. (2016), Steg and Nordlund (2012, 2018), Steg, Bolderdijk, et al. (2014); Steg, Perlaviciute, et al. (2014); Swim et al. (2012).

These framings shape how individuals perceive and process information and act, depending on which goal is dominant.

Further enhancing the TPB, the Theory of Reasoned Goal Pursuit (TRGP) introduces active procurement and approval goals as determinants of behavioral and normative beliefs. These goals influence motivation and can moderate the relationship between attitudes, subjective norms, and motivation. According to TRGP, motivation leads to intention, which ultimately results in behavior (Ajzen & Kruglanski, 2019).

This succinct overview of major theories and models in environmental psychology highlights both their insights and their limitations. While these frameworks offer valuable perspectives on understanding environmental behavior, they predominantly focus on individual human beings, often treating them in isolation. The emphasis is on cognitive processes and personal values as primary drivers of decision-making and behavior. This focus can obscure the broader context within which individuals operate. For instance, one might be aware of the benefits of public transportation but face physical restraints that make it inaccessible. Similarly, a desire to support renewable energy or purchase organic produce can be curtailed by financial constraints. Thus, environmental behavior is influenced by more than just beliefs, attitudes, and values; it is shaped by a complex interplay of factors that these theories might overlook.

By focusing on the individual, the discipline often downplays or disregards systemic issues and conceals its own ideological, hegemonic, and colonial underpinnings, the functions of these theories, as well as the interests they serve (Billig, 2018; Malm, 2025; see also the works of Readsurama Decolonial Editorial Collective comprised of Shahnaaz Suffla, Kopano Ratele, Nick Malherbe, Glenn Adams, and Geetha Reddy). The discipline operates under an assumption of a post-apocalyptic existence—living as if after the end of time rather than within the ongoing crises of the time of the end. This perspective limits its capacity to address the collective and structural dimensions necessary for a comprehensive understanding of environmental behavior and the pressing challenges of our ecological reality. Besides, theories and models of environmental psychology tend to be anthropocentric, emphasizing human agency while neglecting the role of material agency.

Recognizing these limitations, Tabea Scheel and I developed the Modular Interdependency Model (MIM) to situate human environmental behavior within a broader societal context that includes politics, economy, culture, and the ecosphere. MIM expands the list of environmental behavioral determinants beyond intention, attitude, behavior, and personal values to encompass a range of factors, including cultural identity and memory, gender identity, sense of place and belonging, racial and ethnic identity, religious identity, socio-economic status, corporeality, regulatory systems, political systems, political persuasion, technology, lifestyle, and their complex co-constructive interactions and imbrications.

MIM acknowledges that merely providing information is insufficient for behavioral change. Despite being fully aware of the health risks, for instance, many smokers do not quit. Conversely, behavioral changes can occur without direct information provision, such as altering energy consumption patterns through architectural modifications or the automatic reduction in water consumption when aerator or water-saving fixtures are installed in buildings.

Moreover, MIM strives to address cognitive, visceral, as well as imaginative aspects of behavior (Cf. Elaboration Likelihood Model or ELM, as well as fast and slow thinking[2]; Carpenter, 2020; Cheng et al., 2024; Feng et al., 2023; Hedhli & Zourrig, 2023; Kahneman, 2011; Meyers-Levy & Malaviya, 1999; Petty & Cacioppo, 1986; Shahab et al., 2021; Yang & Lee, 2023). It recognizes that behavior is influenced not only by rational deliberation but also by emotional and imaginative engagement. This comprehensive approach aims to capture the full spectrum of influences on environmental behavior, providing a more holistic understanding of the factors that drive individuals and societies to act in environmentally responsible ways. By integrating these diverse elements, MIM seeks to offer a more robust framework for fostering meaningful and sustainable behavioral change.

The model further strives to offer a more comprehensive framework for understanding environmental behavior by incorporating sectoral, geographic, and demographic dimensions. In foregrounding heteronomy, it recognizes the significant roles played by institutions and corporations, acknowledging that agency and responsibility extend beyond the individual. It relies upon the premise that "All things are in the grip of inadequate causation; namely, they are partially determined to act by other, external things" (Lordon, 2014, n.p.). This is critical because individuals may feel

[2] Ironically, models such as the ELM, which are prevalent in marketing research, are applied in contexts like sustainability, even though marketing and sustainability often present challenging reconciliations, much like the tension between class consciousness and positive psychology.

disheartened if their efforts are undermined by the continued negligence or self-interest of others. What is the use of all the lifelong austerity that I impose on myself and my loved ones for environmental reasons when a billionaire's rocket launch can undo that in a fraction of a second? The impact and scale of political and corporate actions far exceed those of individual initiatives, underscoring the need for any model to integrate both individual and collective environmental behaviors.

Unlike previous models, MIM also acknowledges the complex web of intersecting and symbiotic relationships within which humans and other species as holobionts exist. Its multidimensional and decentralized structure aims to move beyond anthropocentrism by positioning individuals and institutions within the ecosphere, emphasizing their interconnectedness with other species and vibrant matter. This holistic approach encourages individuals to recognize their symbiotic relationships and their interdependence as holobionts on other species and material agents within the ecosystem as a crucial step toward behavioral change. This, however, should not lead to a deflection of responsibility. There is a risk that emphasizing external forces may undermine individual responsibility and allow for evading pro-environmental actions. The tension between individual and material agency does not contradict the role of human behavior in driving anthropogenic ecological mutations; instead, it actually highlights the necessity of behavioral change. It calls for "meticulously listening to scientists and putting their results and conclusions into public and democratic discussions, rather than sinking into a geocracy of technological and market-based 'solutions' to 'manage' the entire Earth" (Bonneuil & Fressoz, 2016, n.p.).

Nonetheless, the role of visceral and imaginative dimensions in environmental behavior remains underexplored in MIM, particularly in the context of using satire as a strategy for critical environmental advocacy. As Latour poignantly reflects, "it is hard not to wonder what effect the news we hear every day about the state of the planet has on our mental state. How can we not feel inwardly undone by the anxiety of not knowing how to respond?" (Latour, 2018, n.p.). Satire can serve as a potent experiential tool in advancing critical environmentalism through engaging cultural productions.

While studies on the use of humor, satire, and comedy in environmental advocacy are limited, they often argue that these approaches incite positive emotions like hope. However, I prefer to approach such claims with caution. Recent empirical evidence suggests that classifying emotions based on their certainty, rather than simply as positive or negative, may offer more insights. Emotions can be categorized according to the level of confidence they elicit. For example, emotions such as anger, disgust, contentment, and happiness are associated with a sense of certainty, despite the first two being negative and the latter two positive. Conversely, emotions like worry, shame, surprise, and hope are typically accompanied by a sense of uncertainty. It is this level of confidence that influences the depth of processing.

Humor, in particular, tends to reduce processing depth by eliciting a positive mood that conveys comfort and certainty. This sense of certainty serves as an internal cue that one is already correct and accurate, thereby diminishing the motivation to process a message deeply and effortfully. Thus, while humor can be a powerful tool in environmental advocacy, its impact on emotional processing and message engagement requires nuanced understanding and careful application (Strick, 2021).

In addition to the nuanced understanding of emotions in environmental advocacy, an important consideration is the role of hope. In his *Requiem for a Species* (2010), Clive Hamilton argues that hope can be a barrier to action, describing it as the "unalterable hope that everything will get better and that the worst is not always a sure thing" (Latour, 2017, p. 73). He suggests that to truly engage with the challenges we face, we must first abandon this overly optimistic framing of life. Others have compared our relation to hope to that of a prisoner (Zechariah 9:12) and slave (CAConrad, 2015). Both metaphors are clearly at odds with the cause of emancipatory worlding. In a more recent example (*Last Week Tonight with John Oliver*, season 12, episode 23), John Oliver meets Bernd das Brot in a white void. After John tells him he hosts a late-night show on HBO, Bernd responds he hates comedy, but adds he has a joke for John anyways: "hope." Latour (2017) echoes this sentiment, advocating for what he terms "dis-hoping."

Latour's concept of "dis-hoping" differs from climate doomism or alarmism in that it involves confronting the realities of our situation without the illusion of a quick fix or miraculous cure. He suggests that progress should be reimagined as a sort of retrogression—a different way of experiencing time that does not rely solely on hope. This approach calls for a subtle form of "dis-hoping," which is not synonymous with despair, but rather a rejection of hope as the primary means of engaging with the passage of time. It is about not relying on hope alone to navigate our environmental challenges. Although this idea may initially seem counterintuitive, it invites a more grounded and perhaps more effective engagement with the pressing issues at hand (Latour, 2017, p. 13).

Therefore, a more nuanced approach to understanding and advocating for environmental behavior needs to engage the experiential, visceral, and imaginative as "the right mood or landscape of affect has to be in place, if a set of moral principles is actually to be lived out" (Bennett, 2010, p. xii). Such an approach can indeed benefit from insights drawn from affect theory (Massumi, 2002; see also Ahmed, 2004; Mayes & Center, 2025). This perspective underscores the importance of both affect and emotions, recognizing their distinct but interconnected roles.

Affect refers to the bodily experiences and intensities that are often preverbal and unconscious. This aspect is crucial not only for its experiential and visceral nature but also because it acknowledges and engages with the corporeal and material dimensions of existence. Affect encompasses the immediate, often unarticulated, responses that individuals have to their environment, which can drive behavior in powerful ways. Emotion, on the other hand, is more structured and tied to language and societal norms. It represents or articulates affect in ways that are intelligible within cultural and social contexts. By politicizing and historicizing the visceral and experiential dimension, emotion provides a framework for understanding how these bodily experiences are shaped and given meaning within a given socio-political landscape (Massumi, 2002, p. 27).

Incorporating both affect and emotions into environmental advocacy allows for a richer engagement with the complex ways individuals experience and respond to ecological issues. It recognizes that while emotions provide a narrative and rational structure to our experiences, affect captures the more immediate, embodied responses that often drive action. This dual focus can enhance the effectiveness of advocacy efforts by addressing both the visceral push of affect and the interpretive pull of emotion, fostering a deeper and more holistic understanding of environmental behavior.

The traditional binary between thought and emotion is increasingly seen as inadequate, similar to the historical misunderstandings caused by the soul/body dichotomy. This is partly due to the growing recognition of the pervasive influence of emotions on all forms of psychological processing, as highlighted in Dalgleish's "The Emotional Brain" (Dalgleish, 2004). Effective environmental advocacy, therefore, requires a spectrum of affects and emotions that bridge the corporeal, the socio-political, and the discursive. This spectrum includes concepts such as shame, indignation, radical pessimism (see Bogad, 2016; Douglas, 2025; Gana, 2023; Gros, 2025; Norman & Welchman, 2025; Solnit, 2016; van der Lugt, 2025; Wilderson, 2020; Zekavat, 2025b), "dis-hoping," and grief as a form of political engagement, as advocated by Butler (2004, 2009; Cf. One Thousand Coffins, Bogad, 2016, pp. 32–35). This approach recognizes the complexity and depth of human experience and its influence on behavior.

Rather than relying on optimism or hope, a stance of radical pessimism may provide a more grounded and realistic foundation for advocacy. This perspective favors emancipation over justice as a rather elusive desire, suggesting a shift toward transformative political action that acknowledges and addresses the deeper structural, affective, and emotional dimensions of environmental and socio-political challenges. For one, "The words *truth* and *justice* carry amuletic appeal, the inseparable reminders of a divine overseer" (Bloom & Bloom, 1979, p. 66). And I certainly do not wish to invoke any divine authority or create the illusion of a divine status. For me, exacerbating anthropogenic ecological mutations and degradation and the intersecting issues like class, gender, and coloniality remain to be material, worldly, and reified.

## Rehab

Drunkards are prone to making mistakes. Reflecting on my previous project, I recognize that I was naive in my over-reliance on frameworks such as environmental psychology, and the language of the EU commission and UN's Sustainable Development Goals (SDGs). In retrospect, these approaches were insufficient for addressing the systemic issues at the core of environmental challenges. My main oversight was probably the belief that environmental psychology could be expanded to include these broader systemic concerns. However, psychology has hardly moved forward from "a psychic interiority that the analyst could study in abstraction from its vast ecological context" (Bonneuil & Fressoz, 2016, n.p.). And environmental psychology seems increasingly adept at what it was originally designed to do: unequally distributing risk and opportunity, facilitating subsumption under capitalism, and ignoring differentiated responsibility.

The field's methodological limitations are apparent in its reliance on unrepresentative samples, such as college students, and the unrealistic expectation that deeply ingrained behaviors can be changed by brief interventions like watching a short video or a few pictures or reading a short passage. Besides, researching advertising, marketing, and branding with the aim of optimizing them plays into the very capitalist dynamics that contribute to environmental problems in the first place. This approach obscures other critical factors and discourses, treats individuals as if they exist in the vacuum of laboratory conditions, and ignores the significance of continuous exposure and incremental impact over time. Likewise, experiments studying environmental humor assume unrealistic expectations from humor that pretty much no other discourse or mode of communication including language itself can deliver. They tend to forget that humor can be misunderstood or fall flat, as can any type of discourse.

Positive psychology, particularly its assertions about humor, is another area ripe for critique. As Billig (2018) argues, positive psychology thrives within the "happiness industry," a major profit-making enterprise closely tied to contemporary managerial practices. Historically, happiness was seen as a consequence of wealth, but this has now reversed, with happiness itself becoming a commodified goal. Positive psychology has flourished within this trend, attracting significant financial resources and becoming a growth industry within academia.

In this vein, environmental psychology has fallen into the trap described by Saito, who critiques earlier degrowth theorists for focusing predominantly on rabid consumption and self-limitation among consumers. "Earlier degrowth theorists tended to focus on consumption instead, advising self-limitation to consumers—conserve water and electricity, stop eating meat, buy used goods, share products instead of buying them individually" (Saito, 2024, n.p.). By emphasizing changes in individual values and behaviors, they avoided confronting capitalism directly, particularly in the realm of labor.

Another significant oversight in the initial development of MIM was the exclusion of technology and access to it as a crucial factor. While technological innovations have undeniably contributed to environmental crises, they could be equally pivotal in mitigating and addressing these challenges. However, access to technology is unevenly distributed across different communities, exacerbating existing inequalities. The COVID-19 vaccine distribution serves as a poignant example, highlighting how corporate profit often takes precedence over human life, as financial gain was prioritized over the equitable sharing of vaccine technology through open licensing. This situation illustrates the need for a shift from abstract claims to justice and equity as elusive and variable to more radical approaches to emancipation. Instead of perpetuating technological serfdom and user fealty, as critiqued by thinkers like Yanis Varoufakis (2023), we should consider embracing the concept of the commons in managing and distributing knowledge and technology. This approach would prioritize collective benefit and accessibility, fostering innovation that serves the broader good rather than the narrow interests of profit-driven entities.

I would like to acknowledge another misstep in advocating for resilience, particularly in the context of using humor and satire as tools for coping and adaptation. While they do have affordances for fostering resilience, the larger question is whether resilience is truly the goal we should pursue. Ajay Singh Chaudhary, in *The Exhausted of the Earth: Politics in a Burning World* (Chaudhary, 2024), critiques the concept of resilience as a mechanism that supports the status quo. Chaudhary argues that resilience—similar to absorbing, accommodating, and recovering—is fundamentally passive and politically inert. It relies on traits like optimism, intelligence, and creativity, which, according to Chaudhary, serve as ideological prescriptions that maintain existing structures.

Chaudhary further suggests that for some, resilience perpetuates business-as-usual, buying time for crisis managers. For others, it is exhausting and demands internalization of stressors and crises, thereby hindering political action. It eventually upholds "optimism in a world that is failing; intelligence in knowing it is the best because it is the only possible world; creativity in adapting to that world; a belief system and cohesive life narrative that affirms the world as it is and asserts the value of each and every individual even as it prepares many for mass death" (Chaudhary, 2024, n.p.; see also Carton & Malm, 2024).

Resilience and adaptation are indeed placating, and Chaudhary was not the first one who noticed this. The *People's Agreement of Cochabamba* was adopted in April 2010 at the World People's Conference on Climate Change and the Rights of Mother Earth in Cochabamba, Bolivia. Organized as a response to the perceived failures of international climate negotiations, particularly the 2009 Copenhagen Summit, officially the United Nations Climate Change Conference (COP15), the agreement represents a grassroots, anti-capitalist approach to climate justice. It calls for the recognition of Indigenous knowledge, the rights of nature, and an alternative economic system that prioritizes ecological and social well-being over profit. The document criticizes market-based solutions like carbon trading and advocates for policies that hold developed nations accountable for climate change while supporting the rights of frontline communities. In part, it reads,

> We the peoples have the equal right to be protected from the adverse effects of climate change and reject the notion of adaptation to climate change as understood as a resignation to impacts provoked by the historical emissions of developed countries, which themselves must adapt their modes of life and consumption in the face of this global emergency. We see it as imperative to confront the adverse effects of climate change, and consider adaptation to be a process rather than an imposition, as well as a tool that can serve to help offset those effects, demonstrating that it is possible to achieve harmony with nature under a different model for living. (n.p.)

Instead of adaptation and resilience, Chaudhary advocates for externalization, which he views as the foundation of political subjectivity and power. This involves recognizing and politicizing the structural phenomena that resilience seeks to absorb. The concept of the "Exhausted," as opposed to class or multitude,[3] offers an alternative organizing principle.

> The potential subjectivity of the Exhausted is not about the most abject or the most advanced class position. Exhaustion as an organizing principle is specific. It gives meaning to already existing experiences in, through, and around the extractive circuit. It gives form to "affective disorders" and actual passions. In the ecological conditions of this moment, it can connect radically disparate social locations, identities, and class strata. The possibility of the Exhausted is grounded in material conditions, connects ecological necessity with "good-life" desires long dated and disintegrating, and crystalizes into a genuine political consciousness through recognition and struggle against the partisans of business-as-usual and those who find safety in their shadow. These are people bound together through ordinary struggle but also by a pre-political feeling, an impulse, experienced differently, of the untenable and unbearable nature of their disparate situation. The "restoration" of all those "who fluctuate between madness and suicide" begins by *refusing* to be resilient, refusing to absorb maximal degradation with minimal support. Exhaustion here is the ground of a possibility — the possibility to finally let go, to externalize, to politicize, to return to a sanity that is not resilience, but rebellion. The current social and political uprisings and upheavals as we begin the second decade of the twenty-first century are, from the point of view of the resilience regime, "maladaptive" behavior. From the point of view of exhaustion, they are the beginning of political possibility. (Chaudhary, 2024, n.p.)

[3] Cf. Standing's (2011) notion of the precariat.

The concept of the Exhausted connects disparate social locations and identities through shared experiences of exhaustion and the ecological crises of the moment. This perspective suggests that true political consciousness and action arise not from resilience but from rebellion against the untenable status quo (see also Fressoz, 2024).

Kohei Saito, in *Slow Down: The Degrowth Manifesto* (Saito, 2024), echoes similar concerns, emphasizing the need to shift away from growth-oriented paradigms. Both authors call for a rethinking of resilience as a concept that hinders rather than facilitates genuine political and social transformation. Their critique invites a reconsideration of how we engage with environmental and social challenges, advocating for a politicization of these issues that moves beyond individual resilience to collective action and systemic change. I will also take a more critical approach to resilience, adaptation, and coping both as functions of satire and in environmental policies and strive to expose what they conceal and whose interests they serve.

In the face of aggravating climate crises, a proliferation of so-called adaptation measures has emerged, each more extravagant—and absurd—than the last. These projects, ranging from walling off Antarctica to illuminating the Amazon rainforest at night, are not merely fantastical; they are emblematic of a broader trend in which capitalist markets, particularly carbon credit schemes, co-opt environmental discourse to perpetuate business as usual. Carton and Malm record such initiatives in *The Long Heat* (2025), often with a satirical and incisive tone. These proposals rival even the most outlandish fictions, surpassing Jonathan Swift's *Gulliver's Travels* in their sheer preposterousness. Recall Swift's satirical depiction of the Laputan scientists, who, in their obsession with abstract and impractical inventions, sought to extract sunbeams from cucumbers—a critique of the Enlightenment's misplaced faith in progress. Yet, the contemporary landscape of climate adaptation projects, with their grandiose and often unfeasible schemes, renders Swift's satire almost quaint by comparison.

These projects are not the product of fringe ideation but are instead championed by what might be termed a "deranged scientist species"—a class of researchers, academics, and technocrats embedded within a rarefied institutional machinery that thrives on the production of speculative, market-friendly solutions. This machinery is sustained by unprecedented funding streams, the prestige of academic journals, and the patronage of billionaire philanthropists, whose "generosity" often serves to launder reputations while entrenching existing power structures. Far from being an endangered species, these actors are celebrated for their role in perpetuating the myth of techno-utopian salvation—a myth that allows governments, corporations, and their benefactors to make resplendent yet hollow claims about their commitment to sustainability.

This spectacle partly functions as a mechanism for what David Harvey (2004; see also Riofrancos, 2025) terms "accumulation by dispossession": the enclosure of commons, the appropriation of resources, and the expropriation and exploitation of land and labor, all under the guise of innovation and progress. The proliferation of such projects ensures the persistence of the status quo, offering the illusion of action while deferring meaningful systemic change. In this way, the absurdity of these proposals is not a bug but a feature—a distraction that enables the continued expansion of capitalist markets, even as the planet hurtles toward ecological collapse.

Unlike mitigation, which adopts an etiological approach by targeting the root causes of ecological mutations, adaptation and resilience operate as reactive strategies, addressing only its symptoms. This distinction is critical: adaptation and resilience do not challenge the structural drivers of ecological degradation. Instead, they function as mechanisms for the economization and redistribution of harm and benefits—a process that systematically reinforces existing power hierarchies. By outsourcing the burdens of climate disruption along lines of class, race, gender, and geopolitical division (center/periphery, metropole/subaltern), adaptation and resilience ensure that the costs of environmental collapse are disproportionately borne by the most vulnerable, while the privileged few in the metropole continue to accrue benefits.

Far from delivering on its promises of resilience or emissions reduction, therefore, adaptation often exacerbates the very crises it purports to alleviate. Carbon offset schemes, for instance, frequently enable continued pollution by powerful actors while displacing environmental degradation onto marginalized communities. Carton and Malm (2025, p. 226) note that carbon capture and storage/use technologies are a late value form by which they mean, "in the overshoot conjuncture, capital seeks to press negative emissions technologies into the production of exchange-value. Against it stands any interest in deploying them for the use-value of reversal."

In this way, adaptation becomes a tool of what Nancy Fraser terms "progressive neoliberalism":

> In its U.S. form, progressive neoliberalism is an alliance of mainstream currents of new social movements (feminism, anti-racism, multiculturalism, and LGBTQ rights), on the one side, and high-end "symbolic" and service-based business sectors (Wall Street, Silicon Valley, and Hollywood), on the other. In this alliance, progressive forces are effectively joined with the forces of cognitive capitalism, especially financialization. However unwittingly, the former lend their charisma to the latter. Ideals like diversity and empowerment, which could in principle serve different ends, now gloss policies that have devastated manufacturing and what were once middle-class lives. (Brenner & Fraser, 2017, p. 131)

Adaptation and resilience mark a discourse that masks the perpetuation of inequality and ecological destruction under the guise of pragmatic, market-driven solutions. The result is not emancipation or even multispecies environmental justice but the further entrenchment of a system that prioritizes accumulation over sustainability and power over equity.

In addition to the pitfalls of advocating resilience, I recognize that our empirical aspirations and claims to objectivity should have been tempered. The challenge of validating the Modular Interdependency Model (MIM) aligns with the ontological recommendations offered by Lambros Malafouris in *How Things Shape the Mind* (Malafouris, 2013). Malafouris underscores the paradox that arises when attempting to understand phenomena related to the subjective nature of human experience. He draws on Thomas Nagel's insight that "any shift to greater objectivity — that is, less attachment to a specific viewpoint — does not take us nearer to the real nature of the phenomenon: it takes us further away from it" (Nagel, 1979, p. 174).

To address this paradox, Malafouris advocates for a hunter-gatherer analytic orientation. This approach involves engaging with the world not with the intention of achieving mastery and control by dissecting reality into separate analytic domains, but rather by understanding the complex interconnections and relationships that constitute it. As he puts it, "the aim is not to break things apart into a series of easily manipulated isolated entities, comprehensible in themselves yet incommensurable as a whole, but rather to understand how things are enmeshed and related, and to understand through what mechanisms those linkages are effected" (Malafouris, 2013, p. 53).

A more holistic and integrated approach is therefore necessary for understanding complex systems like environmental behavior. Rather than striving for an illusory objectivity, it may be more productive to embrace the interconnectedness and hybridity of reality, endorsing the subjective and relational dimensions of the phenomena we seek to understand. This shift in orientation could lead to more nuanced and effective insights, particularly in addressing the multifaceted challenges of environmental and socio-political issues.

Despite its limitations, environmental psychology provides some insights that can be effective in understanding certain aspects of ecological mutations and initiating discussions. At the very least, it helps illuminate how sciences can function to maintain and perpetuate ideological and power structures. In other words, it can function as a ploy to expose the workings of ideology and psychological manipulations therein that aim to secure specific interests. Such a critical approach to the discipline can reveal some of the ways through which ideology is instilled. Its mechanisms and models can be co-opted for very different reasons compared to the ones they were initially introduced to serve. They can be repurposed to challenge existing narratives, offering alternative perspectives and fostering critical awareness. They might end up engaging individuals shaped by prevailing ideologies, thereby opening pathways for dialogue and reflection.

Strategic employment of the scientific frameworks that the discipline uses can partly counter skepticism and accusations of lacking empirical rigor, thereby providing a defensible basis for claims and interventions in the scientific discourse. Moreover, a critical approach to environmental psychology can assist in dissecting cultural narratives, identifying what is included or excluded, and understanding how these choices either support or resist certain ideologies.

While these affordances are valuable, they must be considered within the broader context of the discipline's propensity to sideline systemic issues. Environmental psychology often focuses on individual behavior, sometimes at the expense of adequately addressing the more significant impact of governmental, institutional, and corporate actions on environmental degradation. We should not lose sight of the larger picture and comply with the unwillingness to acknowledge and address systemic issues, instead focusing on and shifting blame to individuals for the havoc that governments and corporations wreck on the planet. Maintaining awareness of these systemic factors is crucial for developing a comprehensive approach to tackling ecological mutations and advancing environmental advocacy, and this book aims to show how satire can raise such consciousness and counter strategies that deflect responsibility from corporations, policymakers, and institutions.

A similar rigorous critique is required in using other theories and models that have gained prominence in mainstream research and media. One such theory is the Diffusion of Innovations, developed by Everett Rogers (1983). This theory seeks to explain how, why, and at what rate new ideas and technologies spread through cultures. According to Rogers, the diffusion process involves several key elements: the innovation itself, communication channels, time, and the social system. The theory categorizes adopters into five groups: innovators, early adopters, early majority, late majority, and laggards. The adoption process is influenced by factors such as the perceived attributes of the innovation, the type of decision-making process, and the communication channels used.

With its emphasis on demiurges, the Diffusion of Innovations theory bears some resemblance to Richard Rorty's ideas (Rorty, 1989), which suggest that changing the vocabulary of opinion leaders in society can lead to broader societal change. Rorty, however, was more concerned with defying traditional notions of truth and emphasized the importance of reshaping discourse in the public domain. In contrast, the Diffusion of Innovations theory does not explicitly address these anti-epistemological and anti-representational concerns. It assumes a linear model of influence where you reach out to innovators to change their behavior and then early adaptors, early majority, and late majority will simply follow them. Therefore, there are two presumptions engrained in this theory: first, society is stratified and one can manipulate this stratification in order to advance certain positions and serve certain interests. Second, it accepts that a part of society, which it blatantly called laggards, are to be left behind. The combination of excluding certain strata matched with privileged opinion leaders effectively silences those who do not enjoy the same level of cultural capital, leading to institutional and systemic exclusions and power imbalances. Accordingly, it can perpetuate the status quo in which those who bear the brunt of ecological mutations have no say in the decisions made and resources mobilized to ameliorate the crisis. In fact, this is eerily similar to the logic of colonial and neo-colonial ventures, as discussed in works like Malcom Ferdinand's *A Decolonial Ecology: Thinking from the Caribbean World* (Ferdinand, 2022). The theory's objective to stratify and hierarchize the society poses similar risks to the "environmentalism and the technocentric approach to ecological issues [that] lead[s] to the reinforcement of colonial ruptures passed down from colonization" (Ferdinand, 2022, p. 21).

I, therefore, aim to engage with recent theoretical discussions by scholars such as Donna Haraway, Amitav Ghosh, Dipesh Chakrabarty, Ajay Singh Chaudhary, Ursula Heise through analyzing the implications of their works in multimodal cultural productions unfolding under the culture industry. I partly seek to use their theoretical insights to illuminate the ideological functions of scientific disciplines as the enablers and perpetrators of the status quo, particularly highlighting the role of environmental psychology in doing so.

Departing from the individual-centric focus of psychology, this book will address more systemic issues. By examining satire, it aims to bring to the forefront topics often overlooked in environmental psychology, such as labor, class, and inequality; colonialism, racial injustice, and restitution; gender and heteropatriarchy; as well as ecocentrism, emancipation, and material justice. It will emphasize the concept of Homo sapiens as holobionts, underscoring our symbiotic inter-relationships with other species and material actants. It will strive towards a focus on affect and the ethics of alterity (Arendt, 1998; Levinas, 1969, 1985, 1998). Through this approach, the book aims to foster a deeper understanding of the interconnectedness of human and more-than-human lives and the ethical considerations that arise from these relationships.

## Why Chthulucene

This section briefly explores the rationale behind choosing the term Chthulucene over Anthropocene to describe the current epoch. The Anthropocene, or The Age of Humanity, narrates the tale of humanity's emergence as a dominant planetary force, characterized by rampant exploitation of natural resources and consequent carbon emissions. The epoch features a drastic increase in the power of the human species that brings them to the level of geological forces. However, as Christophe Bonneuil and Jean-Baptiste Fressoz argue in *The Shock of the Anthropocene* (Bonneuil & Fressoz, 2016), this narrative is flawed due to its "geocratic" and teleological framing, which simplifies history into reductive graphs and strips ecological issues of their political dimensions.

Bonneuil and Fressoz contend that the Anthropocene's portrayal of humanity as a monolithic agent obscures the uneven distribution of responsibility and impact. "The official grand narrative of the Anthropocene presents not only a unique view of Earth, of which we should all have the same representation, 'from nowhere', but also a humanity seen as biological entity and geological agent" (Bonneuil & Fressoz, 2016, n.p.). They emphasize that the ecological crisis is largely the result of capitalist, militaristic, and industrial systems, rather than a universal human action. Through their examination of concepts like the Thermocene (energy history), Thanatocene (ecocide), Phagocene (mass consumption), and Capitalocene (capitalism's ecological impact), they unravel how power structures have shaped ecological degradation. They caution that the Anthropocene risks perpetuating a modernist fable, overlooking the historical knowledge of ecological harm and the resistance against it, and advocate for rethinking humanity's relationship with nature to address systemic exploitation and environmental injustice. They conclude that "there is some reason to suspect … that the knowledge and discourse of the Anthropocene may itself form part, perhaps unknowingly, of a hegemonic system for representing the world as a totality to be governed" (Bonneuil & Fressoz, 2016, n.p.).

The use of Anthropocene often implies a homogeneous "We" or "our planet," akin to using "man" to refer to all human beings, which perpetuates undifferentiated and normative assumptions. It "conceals a horizontal homogenization and hides internal hierarchizations on both sides" (Ferdinand, 2022, p. 4). Its claim to universality can be wielded to dismiss legitimate concerns over dominant discriminatory practices in the West. The term is laden with heteropatriarchal biases and colonial and imperial ideologies, obscuring the true causes of the crisis and its socio-political and economic costs to various communities. It fails to acknowledge the disparate resources available to different communities for addressing the crisis, and implicitly reifies regimes of truth and certain knowledges—primarily Western and empirical—while discrediting Indigenous wisdom. Consequently, it shifts the burden of responsibility from those who have caused the crisis to those who are most affected by it. Anthropocene is a concept designed to uphold and perpetuate inequalities and reproduce and maintain power structures under a different guise. There is little to no room for the ***de-humanized*** other under Anthropos.

Latour offers a satirical critique of the Anthropocene in his *Gaia*, emphasizing the notion that the "Anthropos" of the Anthropocene is nothing more than a "dangerous fiction of a universalized agent capable of acting like a single humanity" (Latour, 2017, p. 246). He points out that the Anthropocene, contrary to its name, does not represent an exaggerated form of anthropocentrism where humans are elevated to the status of "Supermen." Instead, he argues that the idea of humans as a unified, universal entity must be dismantled into a multitude of distinct peoples, each with their own conflicting interests and competing territories (Latour, 2017, p. 122).

Latour likens the Anthropocene to the biblical story of Babel after the fall of the great tower, highlighting the fragmentation and diversity of human societies. He underscores that humans are not a monolithic entity that can be universally applied or abstracted from their terrestrial context. By doing so, Latour challenges the Anthropocene's narrative, urging a recognition of the complex and varied realities of human existence and the need to understand these differences within the broader context of ecological and terrestrial history (Latour, 2017, p. 122).

A similar critique is advanced by Amitav Ghosh who, in *Smoke and Ashes* (Ghosh, 2024), delves into historical events to critically examine the concept of the Anthropocene. He highlights how the conquest and colonization of the Americas provided Europeans, particularly the English, with a deep understanding of a form of conflict that relied on structural rather than physical aggression to eliminate native populations. This approach allowed Europeans to rationalize their actions by separating nature from the human domain, thus absolving themselves of responsibility for the spread of disease, which they attributed to natural processes beyond their control. This rationalization facilitated destruction through inaction, marking it as a crucial feature of biopolitical conflict.

In other words, Ghosh implies that the Anthropocene perpetuates the human/nature dichotomy that not only places human beings in opposition to and above nature, but also paves the way for the domination, subjugation, and subordination of certain communities by others. He further explores the agency of non-human entities, such as opium, considering it a historical force in its own right. In doing so, he also takes note of challenges in conceptualizing the interactions between humans and plants, which are heavily influenced by class and power differentials. These difficulties are exacerbated by the lack of a vocabulary that accommodates the agency of more-than-human entities in historical narratives (more on anthropomorphism later).

Ghosh also points to recent climate-intensified events as evidence of the vital, agentive properties of various more-than-human forces—biological, geological, and atmospheric. While these forces may appear to be under human control temporarily, they have the capacity to assert their independence and dominance. This perspective challenges the Anthropocene's narrative by emphasizing the complex interplay of human and more-than-human agency throughout history.

Haraway's (2016) concept of the Chthulucene serves as a compelling alternative to the prevailing frameworks of the Anthropocene, urging a profound rethinking of humanity's role within Earth's multi-species ecology. The term encapsulates Haraway's vision of a symbiotic epoch defined by interconnectedness, mutual care, and the flourishing of multiple species. The Chthulucene challenges anthropocentric views by advocating for a relational ontology that recognizes the deeply entangled existence of humans, nonhumans, and ecosystems amidst ongoing planetary crises. It insists that "the established disorder is not necessary; another world is not only urgently needed, it is possible, but not if we are ensorcelled in despair, cynicism, or optimism, and the belief/disbelief discourse of Progress" (Haraway, 2016, p. 51).

Haraway emphasizes the importance of "staying with the trouble," which involves embracing collaborative survival practices rather than resorting to narratives of domination or simplistic salvation. This approach seeks to foster coexistence and kinship within the intricate web of life, encouraging a shift away from human-centric perspectives towards a more inclusive understanding of our place in the world.

By adopting the term Chthulucene, this book aims to transcend the limitations of the Anthropocene narrative. It departs from stewardship to kinship to align itself with the relational turn in sustainability science, a paradigmatic shift from understanding the world as composed of separate, bounded entities to seeing it as co-constituted by dynamic material, ecological, social, cultural, political, and economical relationships (Haider & Rieser, 2025; Jennings, 2016; Selg & Ventsel, 2020; West et al., 2020). This shift challenges mechanistic and dualistic views and embraces interconnectedness, embodied experiences, continually unfolding co-constitutive processes, and more-than-human and alternative epistemologies. Besides revisiting the ontology of relations and embracing the more-than-human world, this turn emphasizes decolonial thinking, as well as symbolic and affective power. It advocates for an approach that not only acknowledges the complex interplay of agency, power, and responsibility, but also actively engages with the intricate, interdependent relationships that define our ecological reality, thereby seeking to promote an emancipatory interaction among human and more-than-human agents.

## Why Satire

Climate narratives remain by and large apocalyptic. Buell (1995, p. 285) observes, "Apocalypse is the single most powerful master metaphor that the contemporary environmental imagination has at its disposal." In a way, "we create images of doom to avert doom" (p. 295; see also Buell, 2003; Morton, 2010). While utilizing fear appeal in environmental discourse may initially capture attention, overreliance on fear-inducing, apocalyptic narratives in addressing ecological mutations carries several potential drawbacks. And Sachs (2023, p. 91) maintains, "we need … less anxiety and fear."

These doleful and gloomy messages, often somber and overwhelming, can lead to eco-anxiety, depression, despair, apathy, and inertia, making it difficult for individuals to respond effectively to ecological and political challenges. They are emotionally debilitating and risk hampering our ability to imagine alternatives to the status quo and driving us to feel impotent in facing ecological crises. By emphasizing individual responsibility, they risk obscuring the systemic changes that are crucial for meaningful change and fail to adequately acknowledge the disparate roles of different communities in contributing to environmental degradation and climate change, as well as the inequalities in resources available to address and mitigate it (Zekavat & Scheel, 2023).

Moreover, apocalyptic narratives often perpetuate anthropocentrism by attributing transformative power exclusively to Homo sapiens, without recognizing the complex interdependencies between humans, other species, and vibrant matter. By blaming human beings, these narratives imply and uphold their position of mastery over non-human material actants. But perhaps the most significant drawback of apocalyptic narratives is their tendency to reinforce colonial and imperial perspectives. These narratives are frequently framed from the vantage point of those in power, who fear losing their dominance and resources. Consequently, the global catastrophes they depict often underscore threats to the privileged group's way of life, including its ethnic, social, and economic structures. This focus can marginalize or entirely overlook the experiences and realities of other communities, whose environments and livelihoods may have been jeopardized long before the threat of apocalypse in the West (Ferdinand, 2022; Hamouchene & Sandwell, 2023; Heise, 2020). Such narratives conceal socio-economic and health realities of people with non-Western temporalities and genealogies who bear the brunt of ecological mutations (Zekavat, 2025a).

While there are plenty of examples of cli-fi and movies depicting New York or Florida submerged under rising sea levels in the future, depictions of the devastating floods in Pakistan during the summer of 2022 that submerged a third of the country, impacting tens of millions of lives and causing billions of US dollars in damage, remain scarce. Rarely do such narratives concern the global majority, the lives of people who live in shantytowns, the disenfranchised, the unspeakable ecological degradations in Palestine due to settler colonialism, apartheid, and omnicide (Burawoy, 2025a; Burawoy, 2025b; Malm, 2025; Mason et al., 2021).

In "The Planetary Futures of Eco-Apocalypse," Heise (2020) examines the central role that apocalyptic narratives have played in environmental discourse, particularly through the medium of science fiction and speculative literature. Heise observes that these narratives blend religious and secular traditions, creating a "master metaphor" for environmental storytelling. Religious apocalyptic visions, such as those found in the Christian Book of Revelation, depict divine agency as the catalyst for the world's end, followed by eventual redemption. In contrast, secular apocalypses focus on human actions, like the exploitation of nature, highlighting human agency and the potential for mitigation. These narratives aim to galvanize audiences towards change, reminiscent of the American jeremiad's call for social reform. Heise further identifies a dual function of apocalyptic storytelling: fostering proleptic imagination, which anticipates and shapes the future, and generating rhetorical urgency, often invoking a sociocultural sense of crisis.

She traces the evolution of environmental apocalypse conceptualization over time before getting to the waning rhetorical power of apocalyptic narratives. Scholars like Frederick Buell suggest that modern audiences perceive environmental degradation as a chronic crisis rather than an imminent and unfolding catastrophe, leading to a "domestication" of the crisis. This shift risks fostering apathy or acceptance of environmental decline as the norm. Indigenous and postcolonial scholars critique apocalyptic projections as future-oriented fears rooted in privileged perspectives. For many marginalized communities, the apocalyptic conditions imagined are present-day realities shaped by colonialism and systemic injustice. Heise further critiques the commodification of apocalypse in popular media, where repeated disaster depictions can trivialize their intended cautionary impact.

Heise continues to explore alternative narrative strategies, advocating for a juxtaposition of apocalyptic tropes with visions of renewal. She argues for narratives that transcend fear-mongering to inspire hope and actionable change (Cf. Butler, 2024; Sachs, 2023), emphasizing that this shift is crucial for addressing the complex, incremental realities of ecological mutations and environmental degradation.

Her call for narratives that offer renewal echoes Meeker. Joseph W. Meeker's *The Comedy of Survival: Studies in Literary Ecology* (Meeker, 1974). Originally published in 1972, the book explores the intersection of literature, ecology, and the human condition, highlighting the ecological importance of comedy as a strategy for survival. Meeker contrasts tragic and comedic modes, suggesting that tragedy perpetuates human alienation from nature by focusing on individual aspirations and conflicts. In contrast, comedy embodies adaptability, cooperation, and resilience—qualities crucial for ecological balance. He argues that comedy celebrates the possibility of survival and biological renewal in a manner akin to a climax ecosystem, where the survival rights of all species are equally recognized. Rather than altering environments, comedy encourages humans to adapt themselves and their attitudes to restore ecological balance.

Satire emerges as a powerful rhetorical strategy capable of effectively conveying urgency, not only in relation to climate change but also concerning the broader urgencies that accompany it, whether as causes or consequences. According to incongruity theories, the violation of expectations is fundamental to satire. By invoking certain expectations only to subvert them, satire initially adheres to accepted norms that uphold the status quo, creating the impression of business as usual, only to ultimately disrupt them and prompt us to question whether this established disorder is necessary after all. Such violation encourages us to scrutinize our beliefs and the systems that shape them.

Furthermore, the cognitive dimensions of satire significantly enhance its emotional appeal as it challenges the domestication of crises and disrupts the futurity of apocalyptic narratives. While Meeker does not emphasize hope to the extent that Heise does, I advocate for a cautious approach to hope, as I have previously explained the potential benefits of emotions like undefeated despair, radical pessimism, or "dis-hoping" in spurring action. Nonetheless, the importance of both affect and emotion cannot be overlooked. As Latour (2017, p. 23) notes, facts are not "unchallenged arbiter[s]," highlighting the necessity of imaginative and visceral engagement.

The rhetoric of satire is daring in its cognitive and visceral appeal, as it boldly derides, deviates from, and violates established norms. It embodies the figure of the drunkard searching for his keys under the streetlight, despite knowing he left them at home, prompting us to question why he cannot afford a lamp in his own home. What does he have in common with the constable that brings them together in the middle of the night? Why does the constable, a figure of authority responsible for guarding the status quo, initially single out the drunkard for his intoxication, yet fail to question the futility of his search despite his evident inebriation, eventually accepting the situation? Who is the true subject of satire—the drunkard, the constable, or the system the constable represents? Is it hope that drives the drunkard to seek what he has lost in a different place, or is it despair at the prospect of finding them where they were lost? Yet, the drunkard's despair is countered by his relentless pursuit. While the constable aims to uphold law and order, the drunkard is propelled by an indomitable spirit to seek alternatives, not hesitating to mock justice as personified by the constable. Satire, therefore, refuses to project the imminent dangers of exponentially increasing anthropogenic climate mutations into the distant future but instead makes it a material reality of our unfolding present. It abandons the futurity of hope and encourages the hard materiality of life through undefeated despair. It empowers the drunkard to speak up to the constable, abrading his veneer of authority and that of the established disorder he represents.

Satire has also the potential to unleash the demon of shame, known in Greek as Aidos (Αἰδώς), which Lacan (2006, p. 368) describes as "the shame that is sensitive to ridicule, and would bring down the curtain." Vanderheiden and Mayer (2025, p. 41) maintain, "As the challenges of the Anthropocene continue to unfold, the strategic use of shame may prove to be one of the most powerful tools available for achieving meaningful and lasting change." They contend that "shame can … function as a potent resource at individual, organisational, and collective levels, particularly in addressing contemporary challenges such as environmental degradation and social inequality" (p. 42). I argue that satire, as a form of critique, can evoke shame, expose injustices, foster empathy, fuel popular mobilization, and bring to light the contradictions inherent in power structures. In this way, it serves as a discursive strategy for the apoplectic "Exhausted of the Earth" (see also Bogad, 2016; Gros, 2025).

However, it is important to acknowledge that satire can also have negative effects. It has the potential to legitimize oppressive discourses, reinforce social hierarchies, and mask symbolic violence (Campbell & Abdullah, 2023; Marx, 2022; Pérez, 2022; Sienkiewicz & Marx, 2021;

Zekavat, 2026). An example of this dynamic is the Riyadh Comedy Festival, where comedy is manipulated as a tool for soft power by a petro-state implicated in widespread human rights abuses and authoritarianism. By hosting high-profile comedic events, the regime seeks to launder its international image, using humor to deflect attention from its repression and complicity in crimes—both domestically and abroad. In such contexts, satire and comedy risk becoming complicit in "whitewashing" oppression rather than challenging it. Thus, while satire can be a powerful tool for challenging the status quo and advocating for change, it must be employed thoughtfully and critically to avoid inadvertently perpetuating the very issues it seeks to critique (Zekavat & Scheel, 2023). We do not want to laugh *with* the constable and *at* the drunkard.

Until the 1970s, the role of satire as a tool for critique and moral commentary was emphasized as a deontic modality, described as "a moral weapon" by Robert C. Elliott (1960, p. 107; see also Griffin, 1994). Satire compels individuals to speak out against societal vices and wrongs. Juvenal is characterized as the upright individual who views the corruption of his time with horror, his heart filled with anger and frustration. While Horatian satire aims to gently reveal societal pretenses with a flick of laughter, Juvenalian satire seeks to cleanse a corrupt society through the intensity of its disdain. Elliott (1960, p. 115) vividly describes this approach: "the satirist burns, roars, rages, flails as with a sword; the victim cowers, blushes, sweats." Satire, in other words, unfolds a visceral interaction.

In *Satire's Persuasive Voice* (Bloom & Bloom, 1979), Bloom and Bloom argue that satire possesses a "potential for reform," (p. 17) with the ability to "effect a gradual moral reawakening, a reaffirmation of positive social and individual values" (pp. 16–17). They observe that satire, as a persuasive art, shares essential parallels with formal rhetoric, suggesting that its power lies in its capacity to provoke critical thought and inspire change through its compelling and incisive critique. The moral view of satire persists in the twenty-first century when Sachs (2023, p. xviii) argues that dark comedy can help environmental efforts as "a morale boost." Such caustic rhetoric and moral imperative enables environmental satire to defy the domestication of the crisis and call for reform at a systemic level.

The animadverting functions of satire extend beyond affect, emotions, and ethics to encompass significant political and politicizing roles. This is especially crucial because depoliticization can leave environmental movements susceptible to blind spots, such as greenwashing, consumerism, and ignoring racial capitalism and miscarriages of climate justice (see Bresnihan & Millner, 2023; for racial capitalism see Robinson, 2019). Bakhtin (1984) situates satire within the broader context of carnival, where it serves as a critical and transformative force, challenging hierarchical structures and authoritative discourses. The carnivalesque nature of satire disrupts the boundaries between high and low culture, creating a space where societal norms are upended and alternative perspectives are explored. Bakhtin highlights the dialogic nature of satire, which engages multiple voices and perspectives, thereby undermining monologic or authoritarian interpretations of reality. By employing satire, texts resist closure and open up interpretive possibilities, fostering a critical awareness of power dynamics and cultural constructs. These qualities make satire ideal for revisionist projects that do not stop at the cosmetic level but aim to initiate systemic changes as required in addressing anthropogenic ecological mutations.

In *Metahistory: The Historical Imagination in Nineteenth-Century Europe* (1973), Hayden White examines the narrative structures and tropes that underpin historical writing, including satire as a mode of emplotment. White identifies satire as a narrative framework that conveys disillusionment or critique toward human endeavors, often highlighting the futility of progress and exposing the ironies within historical events. This renders environmental satire as an appropriate strategy to advance (the critique of) the discourse of sustainability above and beyond its current prejudices and ingrained interests. Satiric emplotment challenges optimistic or heroic interpretations of history by presenting events as cyclical or unresolved struggles rather than linear paths of improvement. Satire therefore could be an apt strategy to critique the narratives of progress à la Latour and ecomodernism. In White's analysis, satire undermines traditional notions of historical causality and coherence, encouraging readers to approach historical narratives with skepticism and critical detachment. By doing so, satire serves as a tool for questioning the authority of historical accounts and their implicit ideological commitments. This resonates with Latours' critique of the traditional emphasis on causes in social science and philosophy, advocating instead for a focus on the consequences and the networks of actors, both human and more-than-human, that shape events and outcomes.

Therefore, satire has the capacity to (re)politicize climate change by exposing underlying power structures and challenging dominant discourses (Hellmann, 2025). The emotional dimensions of satire also carry political significance. As discussed at the outset of this chapter, conventional responses to fear—fight, flight, freeze, or fawn—prove insufficient in addressing climate change. Predominantly, humanity has responded by freezing, effectively remaining disproportionately inactive. This widespread apathy and sense of helplessness can be explained through Protection Motivation Theory (Rogers, 1975) and the Extended Parallel Process Model (Malecki et al., 2025). According to these frameworks, individuals confronted with fear-inducing messages assess both the severity of the threat and their own susceptibility to it, a process known as threat appraisal. They also evaluate their ability to respond effectively, or coping appraisal, which includes perceptions of response efficacy and self-efficacy. If individuals perceive the threat as severe but doubt their capacity to mitigate it, they engage in fear control—managing their emotional distress rather than taking action. Conversely, if they believe effective action is possible, they engage in danger control, actively addressing the threat. When the perceived threat is high but coping efficacy is low, individuals often experience a sense of overwhelm and powerlessness, leading to inaction or a "freezing" response.

Socio-economic classes that hold power and control and leverage them to advance their vested interests through implementing their beliefs and theories in the real world are often aware of these dynamics and skillfully manipulate them, as seen in culture wars. Politicians and corporations have been known to exploit emotionally charged rhetoric and topics, such as environmental concerns or immigration, to deflect blame from themselves. By pitting environmentally conscious progressives against conservatives, culture wars distract people with debates over individual carbon footprints, such as eating meat or flying, rather than addressing the significant environmental impacts of corporate and policymaker (in)actions. By emphasizing individual responsibility, corporations shift the focus away from their own substantial contributions to environmental degradation. This tactic is evident in how the concept of "carbon footprint" was popularized by BP. Similarly, Coca-Cola's plastic advertisement campaigns shift the responsibility for waste on consumers who do not recycle bottles rather than the corporation that keeps rolling out disposable plastic bottles. This strategy divides people, causing them to fight among themselves and overlook the larger systemic issues at play. While individual behaviors do matter, they pale in comparison to systemic issues underlined with the influence of corporate and policymaker decisions. As in John Oliver's pointed critique of BP's advertising (Oliver, 2022) or Joko Winterscheidt's stark revelations in *The World's Most Dangerous Show* (2023), satire can help expose these tactics, by challenging dominant discourses, encouraging a more unified and politically engaged response to climate change that targets the true sources of environmental harm, and leveraging its emotional appeal to counter blame shifting strategies and reductive pseudo-solutions.

Due to the emotional and epistemological uncertainty it promotes, satire therefore could provide a liberating strategy. Willett and Willett (2019) argue that the poststructuralist critique of power's pervasive influence through norms that shape knowledge and discourse suggests that humor and irony could be powerful catalysts for political change. They propose that since social norms shape cognitive habits, disrupting these norms through ridicule might liberate our thinking, offering a more effective means of transformation than appeals to reason alone (see also Rossing, 2016, 2019). Satire's challenges are not limited to the political sphere and can also be directly wielded for critical environmental advocacy.

There are more ways through which the visceral and cognitive appeal of satire and its unsettling qualities can advance political critique and environmental advocacy side by side. In his exploration of eco-comedy, Takach (2022) highlights its potential as a dynamic tool in the environmental humanities. Eco-comedy is not just a niche interest but a powerful medium that can disarm, reveal truths, engage audiences, and inspire beneficial action. It creates new orientations and visions, uniting and empowering people to take action for environmental causes.

Likewise, Nicole Seymour's *Bad Environmentalism: Irony and Irreverence in the Ecological Age* (Seymour, 2018) challenges the conventional, solemn approaches to environmental advocacy by embracing humor, irony, and irreverence as effective tools for ecological engagement. Seymour critiques the exclusivity and moralism often associated with traditional environmentalism, which can alienate potential allies. She further contends that a comedic stance might be particularly effective at a time when people are experiencing "doomsday fatigue" from an overload of tragic narratives. By examining pop culture, performance art, and queer activism, she illustrates how unconventional mediums and playful tones can raise ecological awareness while fostering inclusivity. Her work rejects the notion of ecological purity and acknowledges human imperfections, encouraging a more accessible and intersectional environmentalism that amplifies marginalized voices. Seymour reimagines environmental activism as a space where imperfection and creativity are not just accepted but celebrated, providing a refreshing critique of eco-morality and guilt-driven narratives (on the political dimensions of guilt as compared to shame—an emotion that can be induced by satire—see Bogad, 2016; Gros, 2025).

Humor further possesses a unique ability to reduce message resistance, even among audiences who initially disagree with the underlying content. It achieves this by acting as a discounting cue, which diminishes the motivation to generate counter-arguments or experience reactance when faced with a message that contradicts one's attitudes. This effect suggests that humor can effectively soften resistance to persuasive environmental messages (Moyer-Gusé et al., 2018; Moyer-Gusé et al., 2019; Nabi et al., 2007; Young, 2008, 2020).

Beyond its communicative efficacy, satire has the potential to challenge anthropocentric paradigms and offer new models based on symbiosis and material vibrancy for alternative modes of coexistence. As Skiveren (2024) argues, humor should be seen not only as an efficient mode of communication but as an ontologically formative practice. It produces spaces of ambiguity from which new perceptual, ethical, or emotional relations might emerge. These relations provide one avenue for accommodating the need for alternative, more effective modes of engagement with the hazardous realities of contemporary ecological crises. Humoristic practices can open pathways into an indeterminate field of radical meanings, relations, emotions, and affects by breaking with expectations, norms, and habits. This productive capacity of humor largely stems from its ability to engage with the challenging realities of ecological crises through a mode of play. By doing so, humor not only communicates messages effectively but also fosters a sense of possibility, offering an alternative way to navigate the complexities of environmental challenges.

## Material Entanglements

A significant limitation in (environmental) psychology is its insufficient engagement with materiality, which results in a failure to fully acknowledge the agency of more-than-human actants and the intricate relationship between individual decisions, behaviors, and the material conditions of life. The discipline's bias is symptomatic of the prevailing political and economic system in the society. Bonneuil and Fressoz (2016) critique the prevailing economic theory for its abstraction from material reality. This theory focuses on goods primarily in terms of their psychological effects—as providers of utility—rather than their material characteristics. Similarly, capital is often viewed not as tangible productive equipment but as assets that generate financial flows. This dematerialization leads to the naturalization of the exponential economic growth characteristic of the Anthropocene by detaching it from any material basis. Such abstraction obscures the real, tangible impacts of economic activities on the environment and limits our understanding of the ecological implications of economic growth.

Reclaiming materiality and engaging materially are therefore crucial steps toward addressing this disconnect. By acknowledging the material dimensions of economic and ecological systems, we can better understand the agency of more-than-human actants and the ways in which human behaviors intersect with the material world. This approach necessitates a shift from viewing the environment and economy through purely psychological or abstract lenses to recognizing the concrete and often complex material transactions that define our ecological reality. Emphasizing material engagement can lead to more holistic and effective strategies for addressing environmental challenges and fostering sustainable practices (Cf. Morton, 2010).

I propose that satire can serve as a potent strategy for advancing the discourse of material engagement and acknowledging the agency of more-than-human actants. Latour (2017, p. 212) calls for a reimagining of the contrast between religion, the secular, and the terrestrial. He describes "the terrestrial" as "*immanence freed of immanentization*," which reactivates "*uncertainty as to the ends.*" Satire inherently relies on and propagates uncertainty, across both emotional and epistemological dimensions.

As discussed above—and as noted by scholars like Strick and Ford (2021)—satire fosters emotional uncertainty, challenging the status quo and prompting reflection. Moreover, it introduces epistemological uncertainty, destabilizing fixed truths and encouraging open-ended inquiry (DiCaglio, 2015; Rossing, 2016, 2019; Willett & Willett, 2019). The etymology of "satire" itself was uncertain for a long time, and its correct etymology points to a "bowl of mixed fruits," a medley of possibilities that claims no exclusive truth. This aligns with the idea of engaging with the world as it is, rather than focusing on the end of history, apocalypse as a total and radical rupture.

By embracing this medley of possibilities, satire acknowledges and engages our terrestrial rootedness in the present, encouraging us to endorse and navigate the complexities of our material and ecological reality. It challenges us to reconsider our place within the intricate web of life, fostering a deeper understanding of the interconnectedness of human and more-than-human actants. Through satire, we can explore and express these complexities, opening up new avenues for dialogue and action in addressing environmental challenges.

A productive ambivalence, both cognitively and emotionally, requires embracing uncertainty and reevaluating our perspectives, particularly through concepts like radical pessimism and dis-hoping. As Latour (2017, p. 2018) suggests,

> we must first of all give up hope—which projects us *from* the present *toward* the future—in order to be able to turn ourselves around—being reoriented by some powerful representation *of* the future *in order to* transform the present. All are trying to shift the eschatology of a too-remote future toward the present, without always being aware that those they are addressing believe themselves to be immune to any eschatology since they have moved to the other side.

This approach involves shifting our focus from a distant eschatology to the present, recognizing that many people feel immune to such narratives because they perceive themselves as having already transcended them. Such a shift has two requirements: first, dis-hoping; second, and acknowledgement and celebration of material entanglement in here and now. In other words, we need to recognize that both Homo sapiens and the institution and artefacts they create are part and parcel of what Morton (2010) calls mesh, or the interconnected, non-hierarchical entanglement of all living and non-living entities, where every being is both dependent on and withdrawn from every other. The mesh rejects binary separations (e.g., nature/culture) and emphasizes a flat ontology of coexistence.

Satire could play a crucial role in fostering this productive ambivalence by both inciting uncertainty emotions and depicting the world's inherent precarity. It encourages us to "give up hope" and instead motivates us to change course. Satire, for instance, can present—rather than *re*present—the future as in *Dr. Strangelove* (see Chap. 2), challenging us to confront the uncertainties of our end times. It offers "*an injunction to rematerialize our belonging to the world*," (Latour, 2017, p. 219) prompting us to keep our feet on the ground, embrace immanence, and address the earthbound materiality of the present. Satire's use of the grotesque and scatological (think of the character of Dr. Strangelove) exemplifies its earthbound materiality, grounding us in reality and liberating us from the suspension "between senseless hope and senseless despair" (Latour, 2017, p. 242; see also Wagner, 2023). Through this grounding, satire encourages a more immediate and tangible engagement with the world, urging transformative action grounded in the present rather than deferred to an uncertain future.

Satire can foreground our material entanglements, but there is a significant caveat. Latour seems to have fallen into the trap of representational logic when he calls upon earthbound scientists to "*represent* [my emphasis] agents who have no other voice and who intervene in the lives of many other agents" (Latour, 2017, p. 253; also see pp. 264–266). The representational fallacy, as discussed by Lambros Malafouris in *How Things Shape the Mind* (Malafouris, 2013), critiques the common tendency in archaeology and related disciplines to explain the relationship between cognition and material culture through representational mechanisms. This approach often regards material culture as merely a vessel or "external storage medium" (Malafouris, 2013, p. 237) for mental representations. Malafouris argues that this perspective overlooks significant phenomena crucial to understanding cognition and material culture. Instead, Malafouris advocates for viewing material culture as a fundamental cognitive resource in its own right. While we can create mental representations of the world, the true efficacy of material culture lies in enabling the mind to operate without constructing such representations. This approach allows for thinking through things in action, without the need for mental replication. Material Engagement Theory posits that the world of material things is its own best representation, negating the necessity for the brain to produce internal copies of what is already present in the external environment.

Malafouris offers an alternative vision of cognition that is inspired by the distributed and compositionally plastic engagement of a potter with clay, rather than the linear architecture of a Turing machine. This view suggests that external resources, such as artifacts and technologies, can have a radically enactive, non-representational cognitive function. Even when external representations serve a representational role, they are also engaged with for what they are, not just for what they stand for.

In exploring the origins of symbolic thinking, Malafouris emphasizes that concepts, such as numbers, can be directly grasped by interacting with the material world, bypassing traditional language-associated brain regions. This highlights a fundamental property of material culture: its ability to act as its own representation. The focus should shift from when humans began to represent one thing with another to when and how they became aware of this process, emphasizing meta-representation or meta-cognition—thinking about thinking.

Thus, Malafouris encourages us to consider how material engagement shapes cognitive processes, challenging the traditional representational logic that often constrains our understanding of the mind-world interaction. This perspective invites a deeper exploration of how material interactions contribute to cognitive development and awareness.

## Anthropomorphism

Orcas have been documented presenting humans with parts of fish, marine mammals, and seaweed in 34 recorded instances as an explicit form of cross-species cooperation, as orcas typically flourish on abundant prey. *The Onion* (2025), a satirical newspaper, responded to this news by publishing American Voices from:

> "Lily Filippo, Sand Sourcer," who reflects, "My parents always told me not to accept food from orcas I don't know."
>
> "Travis Reis, Coral Bleacher," who reacts by exclaiming, "I told them, like, 100 times that I'm off harbor seal."
>
> And "Mohammed Shehadeh, Skydiving Promoter," who maintains, "There's no such thing as free seaweed."

The article underscores interspecies interdependencies while poking fun at anthropocentrism and representational logic. But it speaks of Orcas in human terms, that is to say, it anthropomorphizes them.

Material Engagement Theory encourages a greater appreciation of material agency by emphasizing the role of material culture as an active participant in cognitive processes. However, the challenge of anthropomorphism remains, particularly after the project of the Enlightenment (Horkheimer & Adorno, 2016). Latour (2017, p. 273) maintains that "the fiction resides not in giving water a voice but in believing that one *could get along without* representing it *by a human voice* capable of making itself understood by other humans. … The error would lie in believing it possible to take such interests into account without a human who embodies, *personifies*, *authorizes*, *represents* their interests." The real error is in thinking it possible to account for the interests of more-than-human entities outside of human thought and language. In this view, anthropomorphism is seen as an inevitable aspect of our interactions with the more-than-human world which underscores the complexity of acknowledging material agency while recognizing the limitations imposed by our inherent human perspective.

Anthropomorphism seems to be an inescapable aspect of human perception, but Latour seems to mistake it for representational logic. By mapping human constructs like law and politics onto more-than-human agents such as aquifers, Latour seems to have fallen into the trap of representational fallacy again. On one hand, he argues that non-human agents cannot be "*appropriated* by the interested parties, treated as a substance over which they can claim ownership, and be seized as *property*" (Latour, 2017, p. 274). On the other, the act of imposing anthropocentric institutions like laws and government, as well as discursive representations on these entities inevitably involves appropriating and owning them. This process eerily resembles the initial steps of land appropriation: naming it, or more accurately denying its name and renaming it, to assert control. Representational discourse seems to have served as the supreme instrument of immanentization, "that curious way of simultaneously escaping immanence through a misplaced appeal to transcendence, and escaping transcendence by a too-hasty short-circuit into immanence" (Latour, 2017, p. 274). This exposes a fundamental flaw in Latour's call for a "*representative government*" (Latour, 2017, p. 274) for more-than-human entities. Latour does not take heed of the inherent contradictions, to say the least, of representative governments. Even human representative governments often invariably fail to adequately represent diverse human communities, instead securing the interests of a select few. This shortcoming is even more pronounced when considering more-than-human agents, whose interests are challenging to encapsulate within human frameworks.

Neither representational logic nor representational government appears to be the most effective approach for engaging with the complexities of anthropomorphized more-than-human agency. However, anthropomorphism does not necessarily. Anthropomorphism involves viewing the world in human terms, which is an inherent aspect of human cognition and not necessarily a flaw. Anthropocentrism, on the other hand, involves centering human beings, restricting reality to their perceptions, denying alternative epistemologies, and excluding agency from the more-than-human. Anthropocentrism may allow us to relate to more-than-human entities without necessarily imposing the limitations of anthropocentrism, which places humans at the center of reality. As Bennett (2010, p. xvi) writes, "we need to cultivate a bit of anthropomorphism—the idea that human agency has some echoes in nonhuman nature—to counter the narcissism of humans in charge of the world" as it can reveal isomorphisms between material nature and culture (see also Latour, 2014). By recognizing anthropomorphism as an inexorable part of human interaction with the world, we can appreciate material agency in a way that acknowledges the interconnectedness of humans and more-than-human actants. This understanding encourages us to move beyond traditional representational frameworks and explore more holistic and inclusive ways of engaging with the world.

Malafouris (2013) argues that by adopting a symmetric approach, we excel anthropocentrism and enter a space where conventional boundaries between subjects and objects, or persons and things, become blurred. This blurring prompts significant changes not only in how we perceive ordinary objects but also in the language we use to express these perceptions. While speaking about things as agents might imply an inappropriate ascription of human traits to the non-human, Malafouris (2013, pp. 130–131) contends that the critique of anthropomorphism is flawed in two ways: "First, it fails to recognize the important difference between anthropomorphism and anthropocentrism. Second, it begs the question of material agency—that is, it takes 'human' agency for granted and denies material agency *a priori*." While anthropocentrism is a limiting intellectual habit rooted in Western modernity that should be overcome, anthropomorphism is a biological necessity inherent to human cognition. This characteristic has evolutionary significance, emphasizing the strategic advantage of attributing life-like qualities to objects as a cautious interpretative approach.

Malafouris suggests that anthropomorphism should not be viewed as a failure to overcome a problem but as a central feature of human cognition and material engagement that warrants understanding. Human beings are "the embodied measure of all things, yet we are certainly not the *center* of all things," (p. 132) reflecting a meta-humanistic perspective that recognizes human subjectivity without centering it.

The problematic nature of anthropomorphism arises within the modernist framework that rigidly separates persons and things. The non-anthropocentric orientation of material agency recognizes that while anthropomorphism is a part of this perspective, the concept of material agency transcends it. Material agency is broader and more complex, offering a way to understand the interconnectedness of humans and the material world without the pitfalls of traditional anthropocentric thinking.

A touch of anthropomorphism therefore can indeed serve as a catalyst for a sensibility that perceives the world not as a collection of ontologically distinct categories like subjects and objects, but as a network of variously composed materialities that form confederations. As Bennett (2010) suggests, anthropomorphism can illuminate similarities across categorical divides, highlighting structural parallels between material forms found in nature and those in culture. By doing so, anthropomorphism can reveal isomorphisms—shared structures or forms—that exist across these traditional boundaries. This perspective encourages a view of the world as interconnected and interdependent, where human and more-than-human entities participate in a vibrant and dynamic material existence (see also Secreast, 2023).

Anthropomorphism might even serve as an antidote to what de Waal (1997) calls "anthropodenial," which is the failure to recognize human-like traits in other animals and the animal-like traits in ourselves. De Waal cautions against exaggerating the uniqueness of our species, a tendency that the ancients seemingly did not consider, leaving us without a term until now. Anthropodenial represents an attempt to construct a rigid barrier separating humans from the rest of the animal kingdom, reminiscent of René Descartes' view that humans have souls, whereas animals are mere automatons.

Acknowledging that animals are more akin to relatives than machines makes anthropodenial untenable and anthropomorphism not only inevitable but scientifically acceptable, albeit not in all its forms. The resemblance between humans and animals allows for a more productive use of anthropomorphism, provided it is seen as a tool rather than an objective. The aim "should not be … to find some quality in an animal that is precisely equivalent to an aspect of our own inner lives. Rather, we should use the fact that we are similar to animals to develop ideas we can test" (de Waal, 1997, n.p.). We therefore need a discourse that cherishes the corporeal and grotesque (Bakhtin, 1984), operates on the tensions between body and soul (Critchley, 2002), and has a propensity to dinimish rigid barriers and structures to undermine anthropocentrism while galvanizing anthropomorphism for forging interspecies kinships. In other words, we need satire.

Despite his seeming occasional lapses into representational fallacy, Latour largely challenges the strict separation between subjects and objects. He suggests that what defines scientific expression is not the inanimacy of the objects studied, but rather our limited familiarity with these actors or actants. These inanimate actors require more detailed presentation compared to anthropomorphic characters, with whom we assume a greater familiarity. This perspective emphasizes that the subject does not possess mastery, nor can the object be completely deanimated (Latour, 2017, pp. 49–54).

Latour further argues that the perceived inertia of the material world results from how agency is distributed between causes and consequences. By attributing everything to causes and nothing to consequences—except as passive conduits for effects—we strip the vibrant matter of its agency. To truly engage with materiality, we must reject this reductive operation and recognize the agency inherent in consequences. This effect of deanimation is achieved through a causal narrative, but only retrospectively, after agency has been reassigned across a series of consequences. By acknowledging the active role of consequences, we can more accurately appreciate the dynamic interplay of agency in the material world (Latour, 2017, p. 71).

Latour goes even further by arguing that the analysis of meaning, or semiotics, extends beyond discourse, language, texts, or fictions, as signification is a property inherent to all agents due to their agency. He expands on the concept of agency and signification by introducing the idea of a "zone of common exchange" as an intrinsic property of the world itself, not merely a linguistic or discursive phenomenon. He avers, "signification is a property of all agents, in that they never cease to have agency" as they do not cease to bring their existence from the future into the present, as agents continuously take the risk of affirming their existence or face the possibility of disappearing. "In other words, existence and signification are synonyms. *As long as they are acting, agents signify.*" Their actions and the meanings they generate can be observed, interpreted, and translated into language. This reasoning allows Latour to upend poststructuralist constructivism: This "does not mean that 'every *thing* in the world is merely a matter of discourse' but, rather, that every possibility of discourse is due to the presence of agents in quest of their existence" (Latour, 2017, pp. 69–70). The potential for discourse arises from the presence of material agents actively seeking to assert their existence.

Latour's approach animates matter and effectively dissolves anthropomorphism by emphasizing the inherent agency present in everything that exists. He asserts that the capacity for humans to rearrange their surroundings is not unique but a general property shared by all life forms. On

Earth, nothing is passive; rather, the consequences of actions select the causes that will affect them. This perspective necessitates a heightened awareness of how agency is distributed across different entities. In my analysis of creative works in the following chapters, I will try to show how satire can be used to counter anthropocentrism and animate more-than-human agency in human language mediated works.

By extending intentionality to all agents, Latour argues that anthropomorphism quickly dissolves because it introduces the possibility of non-intentional feedback at every level. This means that each agent, while influencing others, is also being influenced in return. Thus, animation and agency are propagated throughout all interactions. Therefore, Latour challenges the concept of "final cause" as understood in Darwinian terms, where each organism is seen as pursuing its own survival interests. He questions what "final cause" means if it is perpetually interrupted by the equally robust intentions and interests of other organisms. This interconnectedness suggests a dynamic and reciprocal world where agency is not centered around human-like intentionality but is instead a shared attribute of all entities, constantly reshaping the web of life (Latour, 2017, p. 99).

The notion that discourse is materially constructed is indeed compelling and logical. However, its ultimate appropriation into linguistic form by humans raises questions about the extent to which it can be used for material articulation or interspecies dialogue. While discourse originates in material existence, it is susceptible to appropriation. An imperfect analogy can be drawn with the use of English language in colonial contexts: primarily a tool of domination, yet capable of being appropriated and adapted by colonial subjects. Nevertheless, the colonizer retains the ability to reappropriate it at a higher level.

This dynamic is similarly evident in the ingrained biases—such as sexism, heteropatriarchy, classism, and speciesism—that have shaped discourse over centuries. These biases highlight the limitations of discourse as a truly radical medium for change. Despite its material roots, language as a signification system remains a human construct, and anthropomorphism remains a potential pitfall as long as communication relies on it, rather than a radically different medium. While Latour's framing of discourse as a radical tool is inspiring, it is essential to approach it with caution. Aligning with Malafouris's more tempered argument, the key challenge lies in being vigilant against falling into the trap of representational logic. This means acknowledging the potential for appropriation and striving to use discourse in ways that resist reinforcing existing power structures, while remaining open to more inclusive and equitable forms of communication.

## Material Entanglement and the Material Conditions of Life

Material engagement extends beyond acknowledging more-than-human agency; it also involves recognizing the intricate interplay between more-than-human agency and human subjectivity. I will try to show how satire can locate Homo sapiens in the biological and material mesh, while underlining the material conditions of their lives, highlighting how natural, environmental, and socio-economic factors are intertwined and co-constitutive.

Consider, for instance, a coal seam—a layer of coal within the earth's crust that acts as both a source and reservoir for gases like methane. Its discovery and subsequent mining can lead to the establishment of a community or town, shaping its economy, social relations, and legal and political structures. The wealth extracted from the mine continues to influence the economy, with its accumulation tilting material relations within the society surrounding the mine. Such shifts can, in turn, alter power dynamics and societal structures. A tangible example is the Persian Gulf countries, which experienced multiple dramatic transformations following the discovery of oil in the twentieth century. Furthermore, the political and socio-economic regimes regulating a society directly and indirectly influence environmental behavior. Regulations may directly shape behavior, while factors like displacement or immigration can alter a community's sense of belonging or esprit de corps, subsequently affecting its environmental actions.

The material conditions of our existence and the intricate interdependencies between these conditions, human subjectivity, and environmental behavior have been profoundly shaped by capitalism over the past centuries. All satiric works I will analyze are part of the culture industry or the mass production of cultural goods that promote passive consumption and reinforce capitalist ideologies (Horkheimer & Adorno, 2016). At the same time, they seek to criticize the negative environmental impacts of the capitalist mode of production and even undermine the system. This is symptomatic of how capitalism inherently relies on, yet simultaneously undermines, the social and natural conditions necessary for its reproduction. This dynamic gives rise to what can be termed ecological contradictions.

Firstly, the imperative to accumulate capital clashes with the necessity to sustain the social and ecological conditions that facilitate this accumulation. Secondly, capitalism seems to possess a tendency to create and precipitate crises for reinvigorating itself, often through degrading and destroying ecosystems. Capitalism's relentless pursuit of immediate monetary profit often stands in stark contrast to the requirements of the long-term sustainability and viability of these foundational conditions for the continuity of labor, extraction, and accumulation.

As capital accumulation relies on natural conditions that it fails to replenish, it inevitably leads to environmental degradation or incurs escalating costs to maintain environmental equilibrium (Fraser, 2022). Following Marx in *Das Kaital*, Foster (1999, 2000, 2013) calls this disruption in the natural metabolic interaction between humans and the planet a metabolic rift. This dynamic operates not only in human-nonhuman interactions but also profoundly affects human-human relations. Since capitalism does not replenish the resources it exploits, it continually seeks new resources, labor forces, and markets. This relentless pursuit has historically fuelled various exploitative practices, including slavery, colonialism, and imperialism. These practices were partly driven by the need to secure new sources of raw materials, cheap labor, and expand markets for goods, often at the expense of marginalized communities and fragile ecosystems. The extraction of resources and appropriation of lands under colonial and imperial regimes disrupts local economies and societies, leading to long-lasting socio-economic inequalities and environmental degradation. The critical and subversive edge of satire can expose these dynamics and set the agenda for restitution in the sustainability discourse.

In *Slow Down: The Degrowth Manifesto*, Kohei Saito (2024, n.p.) astutely points out that "risk and chance are distributed extremely unequally" among different communities that contribute to and are threatened by climate change. While some view this as unproblematic and believe that capitalism can generate wealth, advance technology, and thereby address both environmental and socio-economic issues, Saito insists that this is a myth. "We usually think of capitalism as something that provides wealth and abundance, but the truth is quite the opposite. Capitalism is a system that functions by producing scarcity."

Capitalism, according to Saito, has historically sustained itself by impoverishing people, beginning with the initial enclosures and continuing to the present day. This leads to the paradox known as the Lauderdale Paradox, which suggests that any increase in private riches occurs only through the diminishment of public wealth. The key distinction between private and public wealth is scarcity. Private wealth is generated by deliberately creating scarcity, dividing up the public wealth that is essential for all. Thus, private wealth can only grow by increasing scarcity, again highlighting a fundamental contradiction in the capitalist system. Parrhesia (παρρησία) or speaking truth to power is recognized as one of the functions of satire that can serve as a weapon for the Exhausted of the Earth to oppose these ingrained power structures that are designed to serve the interests of certain socio-economic classes.

Some argue that individual consciousness and responsibility are sufficient to address the environmental and social issues stemming from capitalism. However, this belief represents a misguided shift from institutional and corporate accountability to individual action. Changes in consumer consciousness are often absorbed by capital due to our entanglement in a commodity-based economy that perpetually seeks growth. Marx (1976; see also Giladi, 2024) referred to this ability of capital to absorb societal changes as "subsumption," a process through which the social relations of production permeate labor itself, a process through which our lifestyles are engulfed by capital.

This relentless entanglement creates a socio-economic reality where even waste and destruction are viewed as business opportunities (think of carbon offsetting and credits), as long as they contribute to the creation of scarcity. Waste and destruction transform abundant resources into scarce commodities, thus maximizing value for capitalists. As Saito (2024, n.p.) points out, "Climate change is, in this sense, a business opportunity like no other. Climate change renders water, farmland, and habitation scarce. As this scarcity rises, demand rises too, until it surpasses supply and provides a prime opportunity for capitalists to reap huge profits."

Even more concerning is the way capital employs artificial scarcity through monetary scarcity caused by debt to solidify its domination. Rabid consumption under capitalism is driven by an endless stimulation of wants, which leads not to wealth but to the accumulation of loans. This burden of debt locks workers into a cycle of obedience, effectively making them pawns of the capitalist system. Here, the entanglement of environmental and socio-economic factors becomes evident as financial capitalism and debt are used to reinforce class structures within society. The recalcitrance of disdaining satire could offer a potential strategy to set the agenda for exposing and partly countering the subsumptive tendencies of capitalism.

This system compels individuals to continuously purchase goods in the pursuit of idealized versions of themselves and their dreams (Cf. Lacan's notion of the Real), driving them to work harder only to consume more.

There is no natural end to this cycle. A consumerist society ensures that the aspirations promised by commodity ownership remain unfulfilled. Scarcity, experienced by consumers as dissatisfaction, is a fundamental principle of capitalism, ensuring that those ensnared within it will never find true satisfaction or happiness. This might be what leaves us perpetually drunk and looking for our keys in the dark, despite the conspicuous futility of the task. Saito's critique echoes the insights of the Frankfurt School, which examined how capitalist societies perpetuate dissatisfaction and control.

Financial capitalism and debt serve not only to consolidate class structures within societies but also to create new and more efficient forms of imperialism and neo-colonialism. Nations that have been historically deprived through colonial exploitation and environmental degradation find themselves taking out loans to adapt and build resilience against these new conditions. However, this debt further entrenches their economic struggles, perpetuating a cycle of dependency and control.

In discussing the transition to a sustainable and just society, Saito (2024, n.p.) observes that "we are living in an era where mutual aid and trust in others has been thoroughly dismantled by the forces of neoliberalism." To rebuild these essential trust relationships, he suggests beginning with face-to-face community building and local municipal politics. In this context, humor and satire can play a crucial role in fostering solidarity and contributing to the restoration of trust within communities. By using humor and satire, individuals can break down barriers, encourage open dialogue, and create a sense of shared understanding and camaraderie. These forms of expression can highlight common challenges, critique existing power structures, and inspire collective action.

While satire can be effective in fostering solidarity and rebuilding trust within communities, it is crucial for it to be applied thoughtfully to avoid replicating or reinforcing existing socio-economic power structures. Psychological mechanisms, which are often used to explain environmental behavior, can sometimes consolidate rather than undermine these structures. Likewise, there is the risk of inadvertently perpetuating the same socio-economic power structures satire seeks to challenge.

Andreas Malm and the Zetkin Collective (2021) argue that organized climate denial has played a significant role in shaping a "post-truth" era. This approach prioritizes personal beliefs over objective realities, utilizing cognitive biases, confirmation bias, and motivated reasoning. These psychological mechanisms were strategically employed by "oil interests" when they first confronted the stark realities of fossil fuels (pp. 296–297; also see Chaudhary, 2024 Ch 4). This history underscores the potential for psychological explanations and narratives to be co-opted in ways that perpetuate existing power dynamics, rather than challenging them.

In essence, psychology often focuses more on helping individuals adapt to the existing world rather than actively seeking to change it. For instance, rather than addressing and altering the underlying causes of stress for employees, it frequently develops techniques to help them cope with increasing stress levels. This approach illustrates how psychology can manipulate humor and other mechanisms to serve capitalism by optimizing profit and accumulation, thereby reinforcing class structures. To quote Latour, in his *Down to Earth: Politics in the New Climatic Regime* (Latour, 2018), "it is not a matter of learning how to repair cognitive deficiencies, but rather of how to live in the same world, share the same culture, face up to the same stakes, perceive a landscape that can be explored in concert. Here we find the habitual vice of epistemology, which consists in attributing to intellectual deficits something that is quite simply a deficit in shared practice" (See also Sparrow, 2019).

As a result, recent decades have seen the rise of ideologies that complicate the response to climate change, if not outright skepticism and denialism, instead of directly addressing the underlying issues. Malm and the Zetkin Collective (2021) outline how the far right approaches the warming world of the 2020s and beyond with concepts such as "palindefence" and "palingenesis," as well as the veneration of "the stock" and the deprecation of "the flow." These ideologies are intertwined with theories like the left climate conspiracy and Cultural Marxism (p. 313). They also note that some segments of the working class may be drawn to "fossil fascism" during a mitigation crisis, as they are as determined as anyone else to maintain their current status and possessions (p. 456).

Ironically, scientific advancements have not succeeded in eroding skepticism and denialism regarding climate change. Instead, we witness right-wing extremists gaining ground by manipulating the sustainability discourse in the West. Malm and the Zetkin Collective (2021) note that the climate movement faces a strategic challenge in trying to appeal to workers as beneficiaries of a just transition—through initiatives like a Green New Deal or a Green Industrial Revolution. The difficulty in selling this message is compounded by waning class consciousness and entrenched system-immanence. These factors may predispose some elements of the working class towards "fossil fascism" during efforts to mitigate climate change. The situation could become even more pronounced in a serious adaptation crisis, where the "wages of whiteness" could take on tangible and material significance, given the fact that those most susceptible to the far right's messaging are often individuals who self-identify as white and male (p. 457).

Thus, implicatory denial can be described as the psychological mechanism underpinning capitalist climate governance. It refers to the complex and constantly evolving strategies used to reconcile the moral and psychological dissonance between awareness of climate issues and the actions taken—or not taken—in response. As noted by Cohen, this denial involves developing a "rich, convoluted and ever-increasing vocabulary" to bridge the gap between knowledge and action, with greenwashing serving as a distinct form of post-truth (qtd. in Malm & the Zetkin Collective, 2021, p. 484). And on the top of this, the very reproduction of fossil capital inherently generates ideologies of denial and other irrational pathologies (p. 486).

Alongside our more-than-human kin and vibrant matter, therefore, Homo sapiens are entangled in the material mesh of existence. At the same time, asymmetrical power structures and ingrained systemic inequalities as in human socio-economic institutions like capitalism impact the material conditions of life of different communities and socio-economic classes. Satire as a part of the culture industry is also regulated by and operates under the same system. Such convoluted material and materialist entanglements require materialist strategies for resistance.

## Materialist Strategies

Addressing the pervasive challenges of anthropogenic ecological mutations and the socio-economic structures that perpetuate them requires comprehensive solutions, each with its own potential drawbacks. Two prominent proposals are degrowth and the principle of common but differentiated responsibilities. In his concluding chapter, Saito (2024, n.p.) observes that because capitalism directly or indirectly draws on natural resources to create value and fuel growth, "we cannot solve a problem triggered by capitalism while still preserving capitalism, as there is no other root cause. We must make a thorough break with capitalism to find a solution to climate change." Natural resources are finite, yet the pursuit of growth is infinite, leading to a fundamental contradiction. To address this ecological dilemma, some have proposed the concept of degrowth.

Degrowth refers to a socio-economic paradigm that advocates for the deliberate downscaling of production and consumption. The goal is to achieve ecological sustainability, social equity, and improved well-being, rather than focusing solely on economic expansion. This approach challenges the traditional growth-oriented model by emphasizing the need to transition towards an economy that values *sumak kawsay* and socio-economic and environmental justice, and seeks to redefine prosperity in terms of quality of life rather than material accumulation.

However, Saito (2024) argues that a notion of "degrowth capitalism" would be paradoxical. The core essence of capitalism is rooted in profit-making, market expansion, displacement, and the extraction of human labor and natural resources. To call for the cessation of these elements—to advocate for deceleration—essentially demands the end of capitalism itself. The idea of eliminating the drive for growth to maximize profits, while simultaneously preserving capitalism, is akin to attempting to draw a round triangle. It is a contradiction, merely a fantasy to which satire can hold up a mirror.

Alongside degrowth, the principle of Common but Differentiated Responsibilities (CBDR) is another proposed solution that acknowledges the unequal responsibilities of nations in addressing environmental degradation, given the historical contributions and varying capacities of different countries. It posits that developed nations, having contributed more significantly to global emissions, should bear a larger share of the burden in mitigation and adaptation efforts. However, this framework is not without its challenges either. Malm and the Zetkin Collective (2021, pp. 492-493) highlight a significant shortcoming of CBDR, tying it to the stark reality that "Whites fly, blacks die" (p. 316). They note that while the United Nations Framework Convention on Climate Change (UNFCCC) recognizes differentiated responsibility, this differentiation is deeply intertwined with the legacy of colonial violence—a fact well understood since the Rio Earth Summit. Assuming full responsibility for averting climate catastrophe carries inherent risks; it confronts those in power with the unresolved injustices of the past. Those who have long ignored and profited from these injustices may resist to accept responsibility for cumulative emissions resulting from historical acts of aggression and material superiority. The fear of where such a reckoning might lead could drive a preference for disavowing the problem, using literal or other tactics to maintain a more comfortable psychic economy.

Beyond degrowth and CBDR, recent strategies have emerged that align more closely with the approach I use in this book. Ecosocialism, rooted in Marxist theory, emphasizes class and class dynamics as central to its frameworks. Class struggle relies on the ideas of Marx and Engels and is echoed in Nixon's (2011) concept of the "environmentalism of the poor." Moving away from class as a conceptual category, Standing (2011) proposes the notion of the precariat, while Hardt and Negri (2000) put forward the concept of the "multitude" as a more fitting strategy for the twenty-first century. This idea suggests a diverse and decentralized collective capable of counteracting global capitalism and addressing contemporary challenges. Following Franz Fanon's insights, Chaudhary (2024) introduces the notion of the "Exhausted of the Earth" to emphasize the lived experiences and struggles of those worn down by systemic exploitation and environmental degradation, advocating for a political movement grounded in their realities. While being informed by Marxian thought, this book also uses the conceptual framework of the Exhausted of the Earth that could open up a space both for intersectionality and interspecies and multispecies discourse.

Chaudhary insists that climate change is inherently political. It should not be seen as a vague, impersonal abstraction or a problem that can be neatly "solved" through universal consensus or technological miracles. Instead, climate change represents a political struggle involving real people, real crises, and real differences, interests, and prospects. It is fundamentally about power; it involves the socioeconomic forces that shape our ecological niche, the capacity of these systems to influence political structures, and the potential political power to redirect our burning world toward unforeseen and improved outcomes. Any effective climate politics therefore must begin with the actual conditions before working its way through the historically situated conflicts that arise. Radical realism in climate politics does not entail discarding all questions of compromise or imperfection. Instead, it starts with today's concrete, interrelated ecological and social conditions, building from inseparable ecological and social necessities. Such climate politics informs my analysis in this book.

He examines the etiology of climate change in the same vein when he quotes Julia Steinberger who writes "the root causes of global environmental changes are coupled social and technical phenomena, not natural ones. The social system that dominates the world is not 'the economy', i.e., a system detached from time and space; but capitalism, i.e., a historically specific class-based system of extraction, transformation, and organisation of flows of commodities for exchange in markets" (n.p.). He insists that the idea that fossil capital would willingly relinquish its existence through a legally ordained agreement or comply with legislative or administrative directives without resorting to the use of state and private violence is unrealistic, to say the least.

Chaudhary therefore advocates for a materialist approach to understanding climate change, a perspective that aligns with Andreas Malm and the Zetkin Collective's (2021, p. 333) assertion that "Ideology … is profoundly material: it exists not so much inside people's heads as in what they do." The discourse of satire engages the material and materialist dimensions of existence to advance a climate politics that aims for emancipatory worlding. Chaudhary explores how this materialist ideology shapes our needs, consumption behaviors, and desires. While we consume more, this consumption is not necessarily fulfilling in terms of desire; instead, our fundamental desire to continue existing is redirected into the logic of capital accumulation. Desire can sometimes reflect choices or manifest as needs; however, to assume that our desires are solely determined by material needs may imply that the concept of 'need' itself is shaped by societal expectations. Desire can also present itself as 'need,' offering more open choices between what is perceived as necessary (Cf. Lordon, 2014).

The political agenda that Chaudhary (2024) envisions encompasses several transformative actions aimed at addressing both ecological and social challenges. This includes decarbonization and a transition to renewable energy sources, along with implementing swift and comprehensive restrictions on fossil fuel extraction. He advocates for the decommodification of essential social goods, recognizing that they are more efficient and effective when publicly provided. Satire is also a social good that is frequently commodified under the culture industry, and I will analyze the implications and extent of such commodification in this book. Moreover, Chaudhary calls for a significant reallocation of resources and sovereign power to the Global South, driven not only by moral obligation but also by rational necessity. This shift accompanies a move away from economic growth paradigms toward "growth agnosticism," focusing on human development and redistribution.

Chaudhary also emphasizes the need to put an end to further dispossession and enclosure, advocating for sustainable, agroecological food production. He proposes greater restrictions and public oversight of capital flows, while simultaneously promoting greater freedom and facilitation for human migration. Underpinning these proposals is a vision of a world characterized by enhanced individual, social, and political security, increased temporal luxury, different material freedoms, and greater human flourishing.

This vision, described as a "minor paradise," seeks to transcend dogmatic debates about economic growth or degrowth and ideological discussions about "real socialism" or "managing capitalism." Instead, the focus shifts to socio-ecological relations, distribution, wealth, and profit. The concept of a sustainable niche addresses the exhaustion inherent in the extractive circuit and acknowledges the diverse desires expressed in its quasi-utopian, yet attainable, fragments. It offers a luxurious reimagining of how to meet socio-ecological necessities.

Chaudhary further emphasizes the importance of affect in addressing climate change. Highlighting the anthropogenic causes of climate change can provoke anger, whereas attributing it to divine or natural causes tends to leave people feeling powerless. "Disgust, anger, rage, hate—these are instructive in the face of today's catastrophe—not the fear which petrifies or the anxiety which seeks the flight to safety" (Chaudhary, 2024, n.p.). In contrast to petrifying fear, indignation encourages people to take risks and engage in tangible action. Satire can incite, multiply, and speak to such indignation; therefore, it can advance the politics of exhaustion by concerting affective rage into coordinated political action. It provides an opportunity for the discursive articulation of pre-political affects into emotions that can externalize and call for a politics of rebellion.

While Chaudhary follows Adorno's skepticism of humor, which has led him to overlook the role of satire, it is worth considering that satire inherently incites indignation (Test, 1991; Zekavat, 2014; also see Zekavat, 2025a). This aligns with the idea of Juvenalian satire, which is known for its biting critique of societal issues. Therefore, satire can be a powerful tool in advancing the politics of exhaustion by channeling anger and indignation into meaningful action and resistance.

However, anger, a certainty emotion, may not align perfectly with advancing the politics of exhaustion. Empirical evidence indicates that uncertainty emotions may be more effective in influencing behavioral change. I am aware that the empirical evidence I am citing originates in psychology, which, as Chaudhary goes in length to show, is not compatible with the politics of exhaustion in being complicit in its counterinsurgent position in crisis management under capitalism with its cruel positivity and mandatory optimism. Yet, these findings can be repurposed to serve the politics of exhaustion.

Satire, unlike the "cruel optimism" critiqued by Lauren Berlant and described by Barbara Ehrenreich in *Smile or Die* (2009), does not enforce an obligatory positivity. Instead, it can effectively channel emotions like love and hatred in a constructive manner. And hatred, particularly from the oppressed, is an inevitable and potentially liberating force. This form of hatred is not personal or pathological but is a structural response to the world's injustices, such as genocide, ethnic cleansing, settler colonialism, fascism, and environmental degradation. Abhorrence, in this context, is essential for dignity and political agency. While loathing can serve as a powerful motivator against injustice, it should not be the sole driver for renewal. It is crucial to neither celebrate nor deny its existence (Miéville, 2022). Satire can play a crucial role in inciting these emotions, motivating activist rigor, and fostering political action. By carefully navigating these emotions, satire becomes an ally in the struggle for emancipation and the advancement of the politics of exhaustion. It helps us narrate "what remains is what cannot be quantified: suffering, hopes, struggles, victories, refusals, and desires" (Ferdinand, 2022, p. 18).

Satire as the politics of exhaustion can also facilitate material engagement with and political struggle against colonial, neocolonial, and imperial dynamics, manifesting in both historical forms and new guises. In *Dismantling Green Colonialism: Energy and Climate Justice in the Arab Region* (2023), Hamza Hamouchene and Katie Sandwell discuss how colonial and imperial enterprises have morphed to "green colonialism" and "green grab." They argue that many renewable energy projects in regions with colonial histories, such as Africa and the Arab world, can be seen as thinly veiled extractivist endeavors. These projects often perpetuate capitalist accumulation, colonialism, and military occupation. They expropriate land and resources without benefiting or acknowledging the rights of local communities and rely on undervalued, disciplined labor. I will show how a seemingly progressive agenda like the European Green Deal doubles down on green colonialism which underscores the necessity for an alternative paradigm that pivots on emancipatory worlding.

Guided by the insights of Malcom Ferdinand in *Decolonial Ecology: Thinking from the Caribbean World* (2022), I aim to challenge the homogenizations, hierarchizations, and apoliticism often found in the Western environmental discourse. I emphasize the significance of ecological and more-than-human concerns within (post)colonial struggles for equality and dignity, which are frequently understated. My core conviction is that "both historical colonization and contemporary structural racism are at the center of destructive ways of inhabiting the Earth" (Ferdinand, 2022, p. 11). Following Ferdinand's lead, I expose how colonial and imperial ideologies perpetuate these harmful practices through military and financial coercion. I renounce colonial inhabitation and acknowledge that "unequal ecological exchanges" have allowed empires and colonizing nations to externalize their environmental burdens outside of their continental territories, transforming their peripheries into plantations, into material and human extensions designed to satisfy the desires of the center (Ferdinand, 2022, p. 44), and seek humane ways for decolonial worlding and a decolonial ecology that "challenges the violence inflicted on humans and non-humans by colonial inhabitation. It is opposed to a way of inhabiting ... which turns the Earth's ecosystems into resources for the enrichment of a few, all while entire populations are kept in conditions of food insecurity" (Ferdinand, 2022, p. 187).

This approach fosters a new understanding of ecology as an engagement with otherness and plurality, which lays the groundwork for a shared world. It suggests that ecology involves establishing a cosmopolitical foundation where humans and more-than-humans coexist, allowing the Earth to become something we hold "in common *without owning it*" (Ferdinand, 2022, p. 19). In this view, the world encompasses nature, the Earth, more-than-humans, and humans, while acknowledging various cosmogonies, qualities, and relational ways of being.

This necessitates a historical, situated, and material approach to ecology. It emphasizes that the starting point for understanding ecology must be grounded, rooted in the world, and expressed from the perspective of beings with bodies, color, flesh, and history. This worldly, material cosmopolitical foundation acknowledges the agency of the vibrant matter while still holding Homo sapiens accountable. It aims to forge alliances across species, recognizing that forming a collective body in the world involves acting with an understanding of our interdependent vulnerabilities and contingencies. This approach seeks to rupture global institutions and economic systems that currently dictate a way of consuming the Earth rather than fostering a way of living together harmoniously. It echoes the boisterousness of an unrepentant drunkard who has relapsed again after yet another rehab of compliance and resilience and cannot help taunting the constable.

Over the next five chapters, I will factor in the impact of the culture industry in the creation, distribution, and commodification of environmental satire and weigh in its workings in cinema, on television and streaming platforms, social media, and in a novel to respond to the question Rabaté (2016, p. 92) raises in asking, "if the Master laughs, can the slave laugh as well?". Depending on the level of its entrenchment and commodification, satire can enable the critique of power structures, renegotiation of social norms, and envisioning alternatives through its cognitive and emotional affordances in inciting uncertainty and uncomfortable emotions like shame, indignation, and undefeated despair. It can function as an externalization strategy for advancing the politics of exhaustion by highlighting shared struggles and interdependencies across different species and vibrant matter in the mesh. Examining environmental satire in *Dr. Strangelove or: How I Learned to Stop Worrying and Love the Bomb*; *Don't Look Up*; *The Late Show with Stephen Colbert*; *Last Week Tonight with John Oliver*; *ZDF Magazin Royale*; *Honest Government Ads*; and *Venomous Lumpsucker* eventually culminates in critiquing the limitations of environmental and multispecies justice, particularly its anthropocentric and institutional biases, and posits emancipatory worlding as a radical alternative. This approach, rooted in decolonial ecology and anti-capitalist principles, seeks to dismantle systemic oppression and promote interspecies commoning. By dethroning the human, embracing material entanglements, and denouncing mastery and exploitation of all sorts, emancipatory worlding offers a vision of liberation for both humans and more-than-humans.

## References


Abrahamse, W. (2019). *Encouraging pro-environmental behavior: What works, what doesn't, and why*. Academic.

Ahmed, S. (2004). *The cultural politics of emotion*. Routledge.

Ajzen, I. (1985). From intentions to actions: A theory of planned behavior. In J. Kuhl & L. Beckmann (Eds.), *Action control: From cognition to behavior* (pp. 11– 39). Springer. https://doi.org/10.1007/978-3-642-69746-3_2

Ajzen, I. (1991). The theory of planned behavior. *Organizational Behavior and Human Decision Processes, 50*(2), 179–211. https://doi.org/10.1016/0749-5978(91)90020-T

Ajzen, I., & Kruglanski, A. W. (2019). Reasoned action in the service of goal pursuit. *Psychological Review, 126*(5), 774–786. https://doi.org/10.1037/rev0000155

Arendt, H. (1998). *The human condition* (2nd ed.). University of Chicago Press.

Bakhtin, M. (1984). *Rabelais and his world*. MIT Press.

Bennett, J. (2010). *Vibrant matter: A political ecology of things*. Duke University Press.

Billig, M. (2018). Positive psychology: Humour and the virtues of negative thinking. In F. Maon, A. Lindgreen, J. Vanhamme, R. J. Angell, & J. Memery (Eds.), *Not all claps and cheers: Humour in business and society relationships* (pp. 3–13). Routledge.

Bloom, E. A., & Bloom, L. D. (1979). *Satire's persuasive voice*. Cornell University Press.

Bogad, L. M. (2016). *Tactical performance: The theory and practice of serious play*. Routledge.

Bonneuil, C., & Fressoz, J.-B. (2016). *The shock of the Anthropocene: The earth, history and us*. Verso Books.

Brenner, J., & Fraser, N. (2017). What is progressive neoliberalism?: A debate. *Dissent, 64*(2), 130. https://doi.org/10.1353/dss.2017.0051

Bresnihan, P., & Millner, N. (2023). *All we want is the Earth: Land, labour and movements beyond environmentalism*. Bristol University Press.

Brick, C., & Lewis, G. L. (2016). Unearthing the "green" personality: Core traits predict environmentally friendly behavior. *Environment and Behavior, 48*(5), 635–658. https://doi.org/10.1177/0013916514554695

Buell, F. (2003). *From apocalypse to way of life: Environmental crisis in the American century*. Routledge.

Buell, L. (1995). *The environmental imagination: Thoreau, nature writing, and the formation of American culture*. Harvard University Press.

Burawoy, M. (2025a). Palestine through a South African lens. *New Left Review, 153*, 133–145. https://newleftreview.org/issues/ii153/articles/michael-burawoy-palestine-through-a-south-african-lens

Burawoy, M. (2025b). Why and how should sociologists speak out on Palestine? *The Sociological Review, 73*(2), 249–260. https://doi.org/10.1177/00380261251321676

Butler, J. (2004). *Precarious life: The powers of mourning and violence*. Verso.

Butler, J. (2009). *Frames of war: When is life grievable?* Verso.

Butler, J. (2024). *Who's afraid of gender?* Farrar, Straus and Giroux.

CAConrad. (2015). Slaves of hope live only for tomorrow. *Poetry Foundation*. https://www.poetryfoundation.org/poems/58067/slaves-of-hope-live-only-for-tomorrow

Campbell, C., & Abdullah, T. (2023). *Race, representation, and satire*. Lexington Books.

Carpenter, C. J. (2020). Elaboration likelihood model. In *The international encyclopedia of media psychology* (pp. 1–14). Wiley.

Carton, W., & Malm, A. (2024). *Overshoot: How the world surrendered to climate breakdown*. Verso Books.

Chaudhary, A. S. (2024). *The exhausted of the Earth: Politics in a burning world*. Penguin.

Cheng, P., Wang, W., & Yang, S. (2024). Doing the right thing: How to persuade travelers to adopt pro-environmental behaviors? An elaboration likelihood model perspective. *Journal of Hospitality and Tourism Management, 59*, 191–209.

Corral-Verdugo, V., Frías-Armenta, M., Tapia-Fonllem, C. O., & Fraijo-Sing, B. S. (2012). Protecting natural resources: Psychological and contextual determinants of freshwater conservation. In S. D. Clayton (Ed.), *The Oxford handbook of en-vironmental and conservation psychology* (pp. 581–597). Oxford University Press. https://doi.org/10.1093/oxfordhb/9780199733026.013.0030

Critchley, S. (2002). *On humour*. Routledge.

Dalgleish, T. (2004). The emotional brain. *Nature Reviews. Neuroscience, 5*(7), 583–589. https://doi.org/10.1038/nrn1432

De Groot, J. I. M., & Thøgersen, J. (2013). Values and pro-environmental behavior. In L. Steg, A. E. van den Berg, & J. I. de Groot (Eds.), *Environmental psychology: An introduction* (pp. 141–152). BPS Blackwell.

De Leeuw, A., Valois, P., Ajzen, I., & Schmidt, P. (2015). Using the theory of planned behavior to identify key beliefs underlying pro-environmental behavior in high- school students: Implications for educational interventions. *Journal of Environ-mental Psychology, 42*, 128–138. https://doi.org/10.1016/j.jenvp.2015.03.005

de Waal, F. B. M. (1997, July). Are we in anthropodenial? In *Discover magazine* (pp. 50–53).

DiCaglio, J. (2015). Ironic ecology. *Interdisciplinary Studies in Literature and Environment, 22*(3), 447–465. https://doi.org/10.1093/isle/isu106

Douglas, R. (2025). *Introduction: Living and wasting time in the world-system. Women in world-literature*. Liverpool University Press.

Ehrenreich, B. (2009). *Smile or die: How positive thinking fooled America and the world*. Granta Books.

Elliott, R. C. (1960). *The power of satire: Magic, ritual, art*. Princeton University Press.

Feng, G. C., Luo, Y., Yu, Z., & Wen, J. (2023). Effects of rhetorical devices on audience responses with online videos: An augmented elaboration likelihood model. *PLoS One, 18*(3), e0282663.

Ferdinand, M. (2022). *Decolonial ecology: Thinking from the Caribbean world*. Polity.

Foster, J. B. (1999). Marx's Theory of Metabolic Rift: Classical Foundations for Environmental Sociology. *American Journal of Sociology, 105*(2), 366–405. https://doi.org/10.1086/210315

Foster, J. B. (2000). *Marx's ecology: Materialism and nature*. New York University Press.

Foster, J. B. (2013). Marx and the rift in the universal metabolism of nature. *Monthly Review, 65*(7), 1. https://doi.org/10.14452/mr-065-07-2013-11_1
Fraser, N. (2022). *Cannibal capitalism: How our system is devouring democracy, care, and the planet—and what we can do about it.* Verso.
Fressoz, J.-B. (2024). *More, more, more: A history of the world's excesses.* Penguin Books.
Gana, N. (2023). *Melancholy acts: Defeat and cultural critique in the Arab world.* Fordham University Press.
Gatersleben, B., & Steg, L. (2013). Affective and symbolic aspects of environmental behavior. In L. Steg, A. E. van den Berg, & J. I. de Groot (Eds.), *Environmental psychology: An introduction* (pp. 165–174). BPS Blackwell.
Ghosh, A. (2024). *Smoke and ashes: Opium's hidden histories.* John Murray.
Giladi, T. (2024). The emergence of Marx's concept of subsumption. *Archiv für Geschichte der Philosophie, 106*(3), 611–631. https://doi.org/10.1515/agph-2022-0074
Griffin, D. (1994). *Satire: A critical reintroduction.* University Press of Kentucky.
Gros, F. (2025). *A philosophy of shame: A revolutionary emotion.* Verso Books.
Haider, L. J., & Rieser, J. P. S. (2025). From reflexive co-production to diffractive co-becoming: Insights from new materialism for sustainability sciences. *Humanities and Social Sciences Communications, 12*(1), 787. https://doi.org/10.1057/s41599-025-05101-6
Hamouchene, H. (2023). The energy transition in North Africa: Neocolonialism again! In H. Hamouchene & K. Sandwell (Eds.), *Dismantling green colonialism: Energy and climate justice in the Arab Region* (1st ed., pp. 29–48). Pluto Press. https://doi.org/10.2307/jj.7583922.7
Hamouchene, H., & Sandwell, K. (Eds.). (2023). *Dismantling green colonialism: Energy and climate justice in the Arab region.* Pluto Press.
Harvey, D. (2004). "The 'new' imperialism: accumulation by dispossession". *Socialist Register.* 40: 63–87.
Haraway, D. (2016). *Staying with the trouble: Making kin in the Chthulucene.* Duke University Press.
Hardt, M., & Negri, A. (2000). *Empire.* Harvard University Press.
Hedhli, K. E., & Zourrig, H. (2023). Dual routes or a one-way to persuasion? The elaboration likelihood model versus the unimodel. *Journal of Marketing Communications, 29*(5), 433–454. https://doi.org/10.1080/13527266.2022.2034033
Heise, U. K. (2020). The planetary futures of eco-apocalypse. In J. Hay (Ed.), *Apocalypse in American literature and culture* (pp. 268–280). Cambridge University Press. https://doi.org/10.1017/9781108663557.021
Hellmann, O. (2025). Can *Rick and Morty* save the planet? Re-politicizing climate change through humor and animation. *Television and New Media, 26*(3), 336–354. https://doi.org/10.1177/15274764241271200

*Holy Bible: King James Version.* (1769/2017). Cambridge University Press. https://www.kingjamesbibleonline.org/

Horkheimer, M., & Adorno, T. W. (2016). *Dialectic of enlightenment.* Verso.

Jennings, B. (2016). Reconceptualizing autonomy: A relational turn in bioethics. *Hastings Center Report, 46*, 11–16. https://doi.org/10.1002/hast.544

Kahneman, D. (2011). *Thinking, fast and slow.* Penguin.

Kals, E., & Müller, M. M. (2012). Emotions and environment. In S. D. Clayton (Ed.), *The Oxford handbook of environmental and conservation psychology* (pp. 128–147). Oxford University Press. https://doi.org/10.1093/oxfordhb/9780199733026.013.0008

Kormos, C., & Gifford, R. (2014). The validity of self-report measures of proenvironmental behaviour: A meta-analytic review. *Journal of Environmental Psychology, 40*, 359–371. https://doi.org/10.1016/j.jenvp.2014.09.003

Lacan, J. (2006). *Écrits: The first complete edition in English.* W. W. Norton & Company.

Latour, B. (2014). Agency at the time of the Anthropocene. *New Literary History, 45*(1), 1–18. https://doi.org/10.1353/nlh.2014.0003

Latour, B. (2017). *Facing Gaia: Eight lectures on the new climatic regime.* Polity.

Latour, B. (2018). *Down to Earth: Politics in the new climatic regime.* Polity.

Levinas, E. (1969). *Totality and infinity: An essay on exteriority.* Duquesne University Press.

Levinas, E. (1985). *Ethics and infinity: Conversations with Philippe Nemo.* Duquesne University Press.

Levinas, E. (1998). *Otherwise than being, or beyond essence.* Duquesne University Press.

Lindenberg, S. (2018). How cues in the environment affect normative behavior. In L. Steg & J. de Groot (Eds.), *Environmental psychology: An introduction* (2nd ed., pp. 144–153). John Wiley & Sons Ltd.. https://doi.org/10.1002/9781119241072.ch15

Lindenberg, S., & Steg, L. (2007). Normative, gain and hedonic goal frames guiding environmental behavior. *Journal of Social Issues, 63*(1), 117–137. https://doi.org/10.1111/j.1540-4560.2007.00499.x

Lordon, F. (2014). *Willing slaves of capital: Spinoza and Marx on desire.* Verso.

Malafouris, L. (2013). *How things shape the mind: A theory of material engagement.* MIT Press.

Malecki, W. P., Schneider-Mayerson, M., Petterson, A., Dobrowolska, M., & Thaker, J. (2025). The role of hope and fear in the impact of climate fiction on climate action intentions: Evidence from India and USA. *Poetics, 108*, 101960. https://doi.org/10.1016/j.poetic.2024.101960

Malm, A. (2025). *The Destruction of Palestine Is the Destruction of the Earth.* United Kingdom: Verso.

Malm, A., & Zetkin Collective. (2021). *White skin, black fuel: On the danger of fossil fascism*. Verso Books.

Marx, K. (1976). *Capital. A critique of political economy*. Penguin.

Marx, N. (2022). Home economics: Sitcom capitalism, conservative comedy, and media conglomeration in post-network television. *Communication, Culture & Critique, 15*(1), 21–35. https://doi.org/10.1093/ccc/tcab065

Mason, M., Fakher Eldin, M., & Dajani, M. (Eds.). (2021). *The untold story of the Golan Heights: Occupation, colonization and Jawlani resistance* (1st ed.). I.B. Tauris.

Massumi, B. (2002). *Parables for the virtual: Movement, affect, sensation*. Duke University Press.

Mayes, E., & Center, E. (2025). Learning with student climate strikers' humour: Towards critical affective climate justice literacies. In C. Russell, P. Chandler, & J. Dillon (Eds.), *Humour and environmental education* (pp. 30–48). Routledge. https://doi.org/10.4324/9781003582120-3

Meeker, J. W. (1974). *The comedy of survival: Studies in literary ecology*. Scribner.

Meyers-Levy, J., & Malaviya, P. (1999). Consumers' processing of persuasive advertisements: An integrative framework of persuasion theories. *Journal of Marketing, 63*(4_suppl1), 45–60. https://doi.org/10.1177/00222429990634s106

Miéville, C. (2022). *Communism, the manifesto, and hate*. Salvage. https://salvage.zone/communism-the-manifesto-and-hate/

Morton, T. (2010). *The ecological thought*. Harvard University Press.

Moyer-Gusé, E., Robinson, M. J., & McKnight, J. (2018). The role of humor in messaging about the MMR vaccine. *Journal of Health Communication, 23*(6), 514–522. https://doi.org/10.1080/10810730.2018.1473533

Moyer-Gusé, E., Tchernev, J. M., & Walther-Martin, W. (2019). The persuasiveness of a humorous environmental narrative combined with an explicit persuasive appeal. *Science Communication, 41*(4), 422–441. https://doi.org/10.1177/1075547019862553

Nabi, R. L., Moyer-Gusé, E., & Byrne, S. (2007). All joking aside: A serious investigation into the persuasive effect of funny social issue messages. *Communication Monographs, 74*(1), 29–54. https://doi.org/10.1080/03637750701196896

Nagel, T. (1979). What is it like to be a bat? In T. Nagel (Ed.), *Mortal questions*. Cambridge University Press.

Nixon, R. (2011). *Slow violence and the environmentalism of the poor*. Harvard University Press. https://doi.org/10.2307/j.ctt2jbsgw

Norman, J., & Welchman, A. (2025). The politics of pessimism. In C. Marshall & S. Shapshay (Eds.), *The new Cambridge companion to schopenhauer*. Cambridge University Press.

Oliver, J. (2022). Carbon offsets: Last week tonight with John Oliver (HBO). YouTube. https://youtu.be/6p8zAbFKpW0

Oliver, J. (Host). (2025). *Last Week Tonight with John Oliver* (Season 12, Episode 23) [TV series episode]. HBO.

Pérez, R. (2022). *The souls of white jokes: How racist humor fuels white supremacy.* NYU Press.

Petty, R. E., & Cacioppo, J. T. (1986). The elaboration likelihood model of persuasion. In *Advances in experimental social psychology* (pp. 123–205). Elsevier.

Pfattheicher, S., et al. (2016). The role of moral emotions in pro-environmental behavior: A literature review. *Environmental Science & Policy, 55*(1), 57–66.

Rabaté, J.-M. (2016). Can you spare a laugh? Lacan, Freud, and Marx on the economy of jokes1. In P. Gherovici & M. Steinkoler (Eds.), *Lacan, psychoanalysis, and comedy* (pp. 82–103). Cambridge University Press.

Riofrancos, T. (2025). *Extraction: The frontiers of green capitalism.* W.W. Norton.

Robinson, C. J. (2019). *Cedric J. Robinson: On racial capitalism, black internationalism, and cultures of resistance.* Pluto Press. https://doi.org/10.2307/j.ctvr0qs8p

Rogers, E. M. (1983). *Diffusion of innovations* (3rd ed.). Free Press of Glencoe.

Rogers, R. W. (1975). A protection motivation theory of fear appeals and attitude change. *The Journal of Psychology, 91*(1), 93–114. https://doi.org/10.1080/00223980.1975.9915803

Rorty, R. (1989). *Contingency, irony, and solidarity.* Cambridge University Press.

Rossing, J. P. (2016). A sense of humor for civic life: Toward a strong defense of humor. *Studies in American Humor, 4*(1), 1–21. https://doi.org/10.5325/studamerhumor.2.1.0001

Rossing, J. P. (2019). An ethics of complicit criticism for postmodern satire. *Studies in American Humor, 5*(1), 13–30. https://doi.org/10.5325/studamerhumor.5.1.0013

Sachs, A. (2023). *Stay cool: Why dark comedy matters in the fight against climate change.* NYU Press. http://www.jstor.org/stable/jj.8784649

Saito, K. (2024). *Slow down: The degrowth manifesto.* Astra Publishing House.

Schultz, P. W., & Kaiser, F. G. (2012). Promoting pro-environmental behavior. In S. D. Clayton (Ed.), *Oxford library of psychology. The Oxford handbook of environ-mental and conservation psychology* (pp. 556–580). Oxford University Press.

Schultz, P. W., Messina, A., Tronu, G., Limas, E. F., Gupta, R., & Estrada, M. (2016). Personalized normative feedback and the moderating role of personal norms: A field experiment to reduce residential water consumption. *Environment and Behavior, 48*(5), 686–710. https://doi.org/10.1177/0013916514553835

Schwartz, S. H. (1992). Universals in the content and structure of values: Theoretical advances and empirical tests in 20 countries. In M. Zanna (Ed.), *Advances in experimental social psychology* (25th ed., pp. 1–65). Academic Press. https://doi.org/10.1016/S0065-2601(08)60281-6

Schwartz, S. H. (1994). Are there universal aspects in the structure and contents of human values? *Journal of Social Issues, 50*(4), 19–45. https://doi.org/10.1111/j.1540-4560.1994.tb01196.x

Secreast, D. (2023). Carson's can of worms: Grotesque satire and abjection in *Rachel Carson's Silent Spring. ISLE: Interdisciplinary Studies in Literature and Environment, 30*(2), 450–469. https://doi.org/10.1093/isle/isab021

Selg, P., & Ventsel, A. (2020). The "relational turn" in the social sciences. In *Introducing relational political analysis. Palgrave studies in relational sociology*. Palgrave Macmillan. https://doi.org/10.1007/978-3-030-48780-5_2

Seymour, N. (2018). *Bad environmentalism: Irony and irreverence in the ecological age*. University of Minnesota Press.

Shahab, M. H., Ghazali, E., & Mohtar, M. (2021). The role of elaboration likelihood model in consumer behaviour research and its extension to new technologies: A review and future research agenda. *International Journal of Consumer Studies, 45*(4), 664–689.

Sienkiewicz, M., & Marx, N. (2021). Appropriating irony: Conservative comedy, Trump-era satire, and the politics of television humor. *Journal of Cinema and Media Studies, 60*(4), 85–108.

Skiveren, N. (2024). Humor as hope? *Environmental Humanities, 16*(2), 441–459. https://doi.org/10.1215/22011919-11150075

Solnit, R. (2016). *Hope in the dark: Untold histories, wild possibilities* (3rd ed.). Haymarket Books.

Sparrow, J. (2019, March 10). Against the new vitalism. *New Socialist*. https://newsocialist.org.uk/against-the-new-vitalism/

Standing, G. (2011). *The precariat: The new dangerous class*. Bloomsbury Academic.

Steg, L., Bolderdijk, J. W., Keizer, K., & Perlaviciute, G. (2014). An integrated frame-work for encouraging pro-environmental behavior: The role of values, situational factors and goals. *Journal of Environmental Psychology, 38*, 104–115. https://doi.org/10.1016/j.jenvp.2014.01.002

Steg, L., & Nordlund, A. (2012). Models to explain environmental behavior. In L. Steg, A. E. van den Berg, & J. I. M. de Groot (Eds.), *Environmental psychology: An introduction* (pp. 185–195). BPS Blackwell.

Steg, L., & Nordlund, A. (2018). Theories to explain environmental behavior. In L. Steg & J. de Groot (Eds.), *Environmental psychology: An introduction* (2nd ed., pp. 217–227). Wiley. https://doi.org/10.1002/9781119241072.ch22

Steg, L., Perlaviciute, G., van der Werff, E., & Lurvink, J. (2014). The significance of hedonic values for environmentally relevant attitudes, preferences, and actions. *Environment and Behavior, 46*(2), 163–192. https://doi.org/10.1177/0013916512454730

Steward, H. (2018). Akrasia. In Routledge eBooks. https://doi.org/10.4324/9780415249126-v003-1

Strick, M. (2021). How humor can promote central-route persuasion: The role of ambivalence. In M. Strick & T. E. Ford (Eds.), *The social psychology of humor* (pp. 3–19). Routledge.

Strick, M., & Ford, T. E. (Eds.). (2021). *The social psychology of humor* (1st ed.). Routledge. https://doi.org/10.4324/9781003042440

Swift, J. (2005). *Gulliver's travels*. Edited with an introduction by Claude Rawson and notes by Ian Higgins. Oxford and New York: Oxford University Press.

Swim, J. K., Markowitz, E. M., & Bloodhart, B. (2012). Psychology and climate change: Beliefs, impacts, and human contributions. In S. D. Clayton (Ed.), *The Oxford handbook of environmental and conservation psychology*. Oxford University Press. https://doi.org/10.1093/oxfordhb/9780199733026.013.0033

Takach, G. (2022). Eco-comedy. *Environmental Humanities, 14*(2), 371–374. https://doi.org/10.1215/22011919-9712445

Test, G. A. (1991). *Satire: Spirit and art*. University Press of Florida.

The Onion. (2025). Orcas bringing humans gifts of food. *The onion*. https://theonion.com/orcas-bringing-humans-gifts-of-food/

Thompson, A. D. (2025, June 30). Daily cartoon: Monday, June 30th [Cartoon]. *The New Yorker*. https://www.newyorker.com/cartoons/daily-cartoon/monday-june-30th-pre-apocalyptic-fiction

Van der Lugt, M. (2025). *Hopeful pessimism*. Princeton University Press.

Vanderheiden, E., & Mayer, C. H. (2025). Investigating shame as a resource in the anthropocene: An emotional psychology perspective. In C. H. Mayer & E. Vanderheiden (Eds.), *International handbook of emotions*. Springer. https://doi.org/10.1007/978-3-031-86449-0_3

Varoufakis, Y. (2023). *Technofeudalism: What killed capitalism*. Penguin.

Wagner, P. (2023). Embracing the environmental grotesque and transforming the climate crisis. *ISLE: Interdisciplinary Studies in Literature and Environment, 30*(4), 911–930. https://doi.org/10.1093/isle/isab086

West, S., Haider, L. J., Stålhammar, S., & Woroniecki, S. (2020). A relational turn for sustainability science? Relational thinking, leverage points and transformations. *Ecosystems and People, 16*(1), 304–325. https://doi.org/10.1080/26395916.2020.1814417

Wilderson, F. B. (2020). *Afropessimism*. Liveright.

Willett, C., & Willett, J. A. (2019). *Uproarious: How feminists and other subversive comics speak truth*. University of Minnesota Press.

Winterscheidt, J. (Host). (2023). *The world's most dangerous show with Joko Winterscheidt* [TV series]. Amazon Prime Video.

Yang, Q., & Lee, Y.-C. (2023). The effect of live streaming commerce quality on customers' purchase intention: Extending the elaboration likelihood model with herd behaviour. *Behaviour & Information Technology, 43*(5), 907–928.

Young, D. G. (2008). The privileged role of the late-night joke: Exploring humor's role in disrupting argument scrutiny. *Media Psychology, 11*(1), 119–142. https://doi.org/10.1080/15213260701837073

Young, D. G. (2020). *Irony and outrage: The polarized landscape of rage, fear, and laughter in the United Sates.* Oxford University Press. https://doi.org/10.1093/poq/nfaa020

Zekavat, M. (2014). A discursive model of satire. *JESELL: Jena Electronic Stud-ies in English Language and Literature.* http://zs.thulb.uni-jena.de/receive/jportal_jparticle_00324293

Zekavat, M. (2025a). Communicating multispecies planetary health through multimodal humour and satire. *Medical Humanities.* https://doi.org/10.1136/medhum-2025-013359

Zekavat, M. (2025b). The paradox of political satire: Navigating critique in culture industry and neoliberal media. *Comedy Studies.* https://doi.org/10.1080/2040610X.2025.2509429

Zekavat, M. (2026). When aporetic satire backfires: Chappelle's transgender jokes gone awry. *Symplokē, 34*, 1–2.

Zekavat, M., & Scheel, T. (2023). *Satire, humor, and environmental crises* (1st ed.). Routledge.

CHAPTER 2

# Learning from the Past: Satire as a Strategy for Crisis Communication

Exploring the history of environmental satire is an undertaking worthy of an entire book, a task that exceeds the scope of this chapter (Cf. Sperrin, 2025). Nonetheless, it is insightful to consider whether fear appeal always dominated crisis communication after the Great Acceleration. In the contemporary world, marked by ecological mutations and socio-economic inequality, it is essential to draw lessons from history to navigate present complexities and build a sustainable future. At the same time, turning to the distant past for solutions to contemporary crises might be misguided, as these challenges are unprecedented. While cultural heritage can provide valuable insights into coexistence with nature and other species, pre-industrial societies are less likely to offer strategies for mitigating climate change, which emerged with the mid-twentieth century's Great Acceleration in its current magnitude (see Malm & Carton, 2024).

Examining historical cautionary tales, nonetheless, can uncover alternative strategies for environmental advocacy. I will analyze Stanley Kubrick's *Dr. Strangelove or: How I Learned to Stop Worrying and Love the Bomb* (1964) to shed light on the emancipatory potentials of satire for political and environmental discourses. Dating back to the early days of the Great Acceleration, this movie uses satire to convey political, economic, military, and environmental complexities. Analyzing this film offers an historical perspective that highlights the potential of satire as a strategy for environmental advocacy.


M. Zekavat, *Leveraging Satire for Environmental Advocacy*,
https://doi.org/10.1007/978-3-032-04838-7_2

While *Dr. Strangelove* primarily focuses on the threat of nuclear warfare during the cold war, its themes of human folly and the catastrophic consequences of unchecked technological advancement resonate with broader environmental issues. Previous studies largely concern the cold war context and fear of nuclear war in their readings (Broderick, 2017). Only, Traub (2018) has cursorily mentioned the movie as a predecessor of cli-fi. Given the environmental impact of militarization and nuclear armament, as well as ongoing contentions around nuclear power as a sustainable source of energy, *Dr. Strangelove* seems as relevant today as it was during the cold war.

By portraying the absurdity of militarization and the potential for anthropogenic ecological disaster, the movie underscores the interconnectedness of environmental concerns with societal and political dynamics. Moreover, its depiction of reckless power pursuits at the expense of rationality serves as a cautionary reminder of the dangers of prioritizing short-term interests over long-term sustainability. This offers valuable insights into the seminality of dis-hoping and acknowledging that we live in the end times in the discourse around both nuclear arsenals and ecological mutations (for a different perspective, see García Mainar, 1994).

*Dr. Strangelove*, directed by Stanley Kubrick and released in 1964, is a satirical black comedy that delivers a biting critique of Cold War tensions and the doctrine of Mutually Assured Destruction (MAD, see Latour, 2017, p. 218). The narrative is set into motion by General Jack D. Ripper, who, consumed by paranoia, unilaterally initiates a nuclear strike against the Soviet Union leading to potentially catastrophic consequences. This act of madness sends the US government into a frantic effort to avert a global catastrophe. As all attempts to recall the bombers fail, the superpowers are forced into an uneasy collaboration. The Soviet Union, alerted to the incoming American B-52s, attempts to shoot them down, but one plane manages to evade their defenses. After the bomber is severely damaged and losing fuel, the eccentric pilot redirects the attack to a nearby Soviet missile base. As the damaged aircraft struggles to deliver its nuclear payload, the film satirizes a familiar narrative: the B-52 crew, now on a suicide mission, are portrayed as heroic and efficient, driven by an idealized form of patriotism. The film's artistic techniques create a tension between emotional engagement and the horrific reality of nuclear war. This Strangelove Effect is embodied by characters like General Turgidson, who exhibits both childlike excitement and horror at the crew's impending success, and the naive pilot who manually opens the bomb doors and rides an H-bomb to the ground, waving his cowboy hat.

At the heart of the crisis is President Merkin Muffley, who, alongside his advisors, including the erratic ex-Nazi scientist Dr. Strangelove, must navigate the chaos and absurdity of the situation. The film deftly employs satire to lampoon the bureaucratic incompetence and ideological extremism that underpin the nuclear arms race. Kubrick's incisive social commentary highlights the absurdity inherent in the reliance on nuclear deterrence and the perilous nature of global security frameworks (Broderick, 1993).

The film's dark humor and sharp critique of political and military hubris are underscored by iconic performances, particularly Peter Sellers, who masterfully portrays multiple roles, including that of President Muffley and Dr. Strangelove. With its enduring relevance, *Dr. Strangelove* remains a seminal work in cinematic satire, offering a timeless reflection on the dangers of unchecked power and the human propensity for self-destruction. Its exploration of these themes continues to resonate, making it a critical prism through which to view contemporary environmental and geopolitical crises.

## Critique of Technological Fixes and Ecomodernism

The catastrophic consequences of unchecked technological advancement in the movie resonate with the critique of ecomodernism. Ecomodernism puts its hopes on technological development to protect nature and simultaneously improve the quality of life for human beings by decoupling economic growth from environmental impacts. *An Ecomodernist Manifesto* maintains, "A good Anthropocene demands that humans use their growing social, economic, and technological powers to make life better for people, stabilize the climate, and protect the natural world" (n.p.). Ecomodernism, therefore, is essentially anthropocentric and technocentric, emphasizing that it is up to Homo sapiens to stabilize and protect nature through "Intensifying many human activities—particularly farming, energy extraction, forestry, and settlement—so that they use less land and interfere less with the natural world" (n.p.). For ecomodernists, this "is the key to decoupling human development from environmental impacts. These socioeconomic and technological processes are central to economic modernization and environmental protection. Together they allow people to mitigate climate change, to spare nature, and to alleviate global poverty" (n.p.). Instead of tempering our impact, therefore, they argue that we can minimize our impact on nature through further developing methods and technologies allowing us to exploit it more efficiently. The writers of the *Manifesto* "affirm one long-standing environmental ideal, that humanity must shrink its impacts on the environment to make more room for nature, while ... [they] reject another, that human societies must harmonize with nature to avoid economic and ecological collapse" (n.p.; also see Huber, 2022; Saito, 2024).

Ecomodernist propositions might sound self-contradictory. Degrowth proposes a different solution for tackling ecological problems, and primarily climate change. It is critical of (fetishizing) the concept of growth as in indicators like Gross Domestic Product (GDP). While unbridled neoliberal and capitalist development, that directly or indirectly draws on natural resources to fuel unlimited growth, jeopardizes ecosystems, multispecies planetary health, and environmental justice,

> advocates of degrowth call for an end to the fetishization of growth in contemporary society, a reduction in energy use and material throughputs in the Global North, and a globally just distribution of wealth and resources. This is necessary, they claim, to phase out fossil fuels, regenerate the planet's damaged ecosystems and attain a decent quality of life for all. Such a programme would require reductions in resource and energy use for many in the Global North, yet this need not lead to asceticism. Rather, its supporters argue it would enable communal luxury within ecological limits. Though disagreements exist within the degrowth movement, most of its adherents envision a future where food production is localized, people have democratic control over issues that affect them, renewable energy infrastructure is decentralized and collectively owned, and public transportation is commonplace. (Heron, 2022, n.p.)[1]

*Dr. Strangelove* offers a sharp critique of the belief that technical investment and advancements, particularly in military technologies, can enhance global security. This notion, embraced by some technocrats and politicians, posits that nuclear arsenals serve as deterrents, intended to intimidate adversaries and terrorists and dissuade bad actors on the global stage from making harmful decisions. However, the numerous wars, acts of violence, and genocides that have unfolded since the advent of the arms race

[1] For an extensive critique of ecomodernism see Barca (2024), Buck (2019), Caradonna et al. (2015), Chaudhary (2024), Crist (2016), Latour (2015), Robbins and Moore (2015), Saito (2024); "What is Modern in Ecomodernism," 2015.

starkly contrast this belief, demonstrating the inefficacy of weaponization and militarization in fostering peace.

The film presents a world where both technology and the users' skills in manipulating it are so advanced that they inevitably trigger a doomsday scenario. Rather than making the world a safer place, these technological advancements ensure and expedite absolute annihilation (Maguire, 2018). Kubrick's portrayal underscores the paradoxical nature of such advancements: while intended to secure peace, they instead create a precarious balance that can easily tip towards disaster. Through its satirical outlook, *Dr. Strangelove* exposes the folly of relying on technological prowess to solve complex geopolitical issues, emphasizing that these advancements, instead of safeguarding humanity, can lead to its ultimate destruction. This critique invites reflection on contemporary over-reliance on technology to address global challenges, urging a reevaluation of strategies that prioritize short-term security or political gain over long-term sustainability and peace.

The titular character serves as a metaphor for the war machine, embodying the fusion of military technology and madness. Formerly known as *Merkwürdigichliebe* before becoming a US citizen, his name, Strange Love, suggests a distortion of love and sexuality, embodying the death instinct (Stillman, 2008). As a technocrat, Dr. Strangelove represents the intellectual force behind the nuclear war apparatus. He is a deranged scientist, symbolizing science and technology gone astray. He is part human, part machine, with his wheelchair integral to his mobility, making him inseparable from it (see Haraway, 1985). His arm operates autonomously, adhering to machinic principles reminiscent of Henri Bergson, and even attempts to strangle him, illustrating its self-destructive nature. This arm also involuntarily performs a *Heil Hitler* salute while Strangelove misaddresses the President as *Mein Führer*, linking his technocratic brilliance to its Nazi roots, perhaps the same technocratic brilliance that compels Elon Musk to Nazi salutes and draws him to AfD. In the film's final moments, amidst the chaos of nuclear war, Strangelove rises from his wheelchair, exclaiming, "Mein Führer, I can walk," symbolizing his rebirth within the warped logic of nuclear devastation (Simon, 2003).

Kubrick's critique extends beyond the dangers of overreliance on technology to solve the very problems it has created. The film also offers insights into the ontological approach underpinning ecomodernism and technological fixes. The doomsday machine in the film is engineered to self-trigger, with no possibility of cancellation, under the assumption that making the process impersonal and irreversible ensures optimal deterrence. However, this assumption proves tragically flawed when a single deranged individual sets off a chain of events leading to global annihilation.

This scenario reflects a broader human tendency to displace agency, thereby avoiding responsibility for making crucial life-altering decisions. In the context of ecological mutations, some deny that human activities are the primary cause of climate change, thus shirking accountability. Others persist in complacency, insisting that their policies, decisions, actions, and lifestyles are unrelated to climate change, thereby resisting any impetus for change. This displacement of agency not only absolves individuals from taking responsibility but also delegates the task of addressing climate issues to technology.

Placing undue faith in technology, which has been a significant contributor to exacerbating ecological mutations, can lead to potentially catastrophic outcomes. By highlighting this misplaced trust, *Dr. Strangelove* serves as a cautionary tale about the dangers of assuming that technological advancements alone can avert disaster. The film's critique of agency displacement reaches a crescendo in the iconic scene where Major Kong, the quintessential American cowboy, rides a nuclear bomb to its target. His ecstatic rodeo-style celebration, screaming "Aaaaaa hoooo! Waaaaa hooooo!" while he rides the bomb in its falling arc waving his hat over his head, celebrating his success, illustrates the spiraling madness of surrendering human agency to technology.

This moment encapsulates the absurdity and peril of relinquishing control to machines, as it underscores the human capacity for self-destructive behavior when responsibility is abdicated. The film suggests that genuine security and sustainability cannot be achieved through technological fixes alone; rather, they require a fundamental reevaluation of collective actions and attitudes toward the planet. By satirizing the blind faith in technological solutions, *Dr. Strangelove* calls for a more responsible and reflective approach to global challenges.

In an era where ecomodernism often champions technology as the panacea for environmental crises, the film's message remains strikingly relevant. It cautions against the allure of simple technological fixes and underscores the importance of acknowledging and addressing the human behaviors and systems that contribute to such crises. Ultimately, *Dr. Strangelove* urges a reassessment of our over-reliance on technology.

The technology itself does not fail; rather, it is our misplaced (counter) religious faith in it that leads to catastrophic consequences. Despite witnessing the dire outcomes of their policies, the politicians in the war room fail to learn any lessons and double down on seeking solutions rooted in a pragmatic Malthusian morality. This is evident in the absurd plans proposed by Dr. Strangelove in the wake of the catastrophe for saving humanity and repopulating the Earth.

Malthusian economics, based on the theories of Thomas Robert Malthus (1766–1834), emphasizes the relationship between population growth and economic stability. In his "Essay on the Principle of Population" (1798), Malthus argues that while population increases geometrically, food production only grows arithmetically. This discrepancy, he contended, would inevitably result in shortages, famine, and societal collapse unless curbed by natural limits or preventive measures. Malthusian population policies thus advocate controlling population growth to prevent economic strain and resource depletion, which can include: moral restraint or encouraging delayed marriage and sexual abstinence until individuals can support a family; preventive checks promoting contraceptive use and family planning to limit births; and positive checks or recognizing that natural phenomena such as disease, famine, and war reduce population when it exceeds carrying capacity. Malthus's ideas have influenced debates on population control, sustainable development, and environmental economics (see Hardin, 1968).

The Malthusian morality surfaces in the arguments of characters like Ripper and Turgidson. Turgidson's acceptance of positive checks is clear when he discusses the potential loss of life: "Mr. President, we are rapidly approaching a moment of truth both for ourselves as human beings and for the life of our nation. Now, the truth is not always a pleasant thing, but it is necessary now to make a choice, to choose between two admittedly regrettable, but nevertheless, distinguishable post-war environments: one where you got twenty million people killed, and the other where you got a hundred and fifty million people killed." President Muffley retorts, "You're talking about mass murder, General, not war." Turgidson responds, "Mr. President, I'm not saying we wouldn't get our hair mussed. But I do say ... no more than ten to twenty million killed, tops." This cold-blooded discussion of tens of millions of lives eerily resembles contemporary discourse regarding climate change-induced displacement as another form of positive check on population.

Politicians, particularly in wealthier nations, often dehumanize displaced individuals by reducing them to mere statistics, overlooking the profound sense of loss and tragedy associated with displacement. Instead, they exploit xenophobia and incendiary anti-migrant rhetoric to stoke fear and advance certain political agendas. The Malthusian discourse on balancing resources and population is frequently summoned to support xenophobic policies, echoing the film's critique of dehumanization and reliance on reductive solutions to complex global challenges. This reflection underlines the dangers of allowing quantitative assessments to overshadow the lived human experiences deeply impacted by global crises, urging a more compassionate and comprehensive approach to addressing these issues.

Kubrick's *Dr. Strangelove* serves as a trenchant critique of the cost-benefit analysis that underpins Malthusian as much as capitalist and neoliberal economies, extending this critique to the regulation of human life and death in wartime scenarios. Characters like Ripper and Turgidson embody this mindset, advocating for an aggressive preemptive strike on Russian targets to curtail further damage. Turgidson's familiar argument—that launching an all-out attack could catch the Russians off-guard and minimize American casualties—reflects a cold calculation of acceptable losses:

> … if … we were to immediately launch an all-out and coordinated attack on all their airfields and missile bases we'd stand a damn good chance of catching 'em with their pants down. Hell, we got a five to one missile superiority as it is. We could easily assign three missiles to every target, and still have a very effective reserve force for any other contingency. … an unofficial study which we undertook of this eventuality, indicated that we would destroy ninety percent of their nuclear capabilities. We would therefore prevail, and suffer only modest and acceptable civilian casualties from their remaining force which would be badly damaged and uncoordinated.

The cost-benefit analysis that drives militarism resonates with unchecked economic growth—a major contributor to contemporary environmental urgency—and its echo in ecomodernist thought, which advocates decoupling development from environmental impact through intensified human intervention.

Turgidson's logic parallels the flawed rationale of intensifying human activities to stabilize nature, a strategy that risks exacerbating the very disruptions it seeks to mitigate. Intensifying human activities is likely to backfire, much like escalating military tensions for peace to prevail. Just as fear mongering against the USSR and communism justifies alarmist military strategies, leveraging fear in environmental politics can open the door to all sorts of abuses. In the end, it is ironically American soldiers and citizens who bear the brunt of these actions. The deterrence war against the Soviet Union and communism is portrayed as a doomed endeavor even before it begins, paralleling the futile measures often adopted to fostering climate resilience and adaptation. As Chaudhary (2024) notes, such measures are often a lost battle from the outset.

In the film, the fear-driven approach culminates in a farcical dialogue between the US President and Russian Premier Dimitri, with whom the President absurdly expresses solidarity: "Well, we'll keep our fingers crossed [hoping your anti-aircraft forces can take the US B52 bomber down], Dimitri, and remember, there's just one thing, we are all in this together. We're right behind you, Dimitri. We're with you all the way." This statement starkly contrasts underlying anti-communist and anti-Soviet agenda, highlighting the absurdity of adversarial postures (Deutsch, 2016; this might remind one of the contemporary dynamics of the Trump Administration with the Kremlin under Putin).

Kubrick's film underscores the dangers of framing complex issues through a lens of simplistic cost-benefit analysis. By doing so, it warns of the pitfalls of merely relying on quantitative assessments that overlook the lived experiences essential to addressing these challenges. *Dr. Strangelove* serves as a cautionary tale about the perils of allowing economic and military calculations to overshadow moral and ethical considerations (Maland, 1979). It calls for a reevaluation of strategies that prioritize growth and dominance at the expense of human and environmental well-being, urging a shift towards holistic and collaborative approaches that recognize the interconnectedness of global systems.

But all attempts fail to abort the crisis and ultimate Malthusian positive checks are unleashed after the dropping of the bomb triggers the doomsday machine. After it becomes clear that everything is lost, Strangelove steps in to unveil the plan B. Instead of aborting disaster, we need to adapt to the new reality and foster resilience. Strangelove avers, "Mr. President, I would not rule out the chance to preserve a nucleus of human specimens. It would be quite easy … heh heh … [*rolls forward into the light*] at the bottom of ah … some of our deeper mineshafts. The radioactivity would never penetrate a mine some thousands of feet deep. And in a matter of weeks, sufficient improvements in dwelling space could easily be provided." After he suggests to the President that people have to live underground for about 100 years (compare my discussion of seasteading in *Venomous Lumpsucker* in Chap. 6), Strangelove continues, "It would not be difficult *mein Führer!* Nuclear reactors could, heh … I'm sorry. Mr. President. Nuclear reactors could provide power almost indefinitely. Greenhouses could maintain plantlife. Animals could be bred and *slaughtered.* A quick survey would have to be made of all the available mine sites in the country. But I would guess … that ah, dwelling space for several hundred thousands of our people could easily be provided." The sheer insanity of his absurd proposal fades away when one remembers some actual proposed measures to adapt to the brave new world where extreme weather events are a part of normal life (Carton & Malm, 2025). These measures and projects are being proposed and valorized when we still have a chance to abort the crisis from happening.

The notion of resilience, therefore, needs to be treated with skepticism. Chaudhary (2024) argues that for some, resilience serves as the fundamental principle that allows for the continuation of business-as-usual, essentially enabling crisis managers to delay dealing with pressing issues. Opposing resilience and the tendency to internalize stressors and structural challenges, therefore, is a viable approach to climate realism that relies on the externalization and politicization of climate change and intersecting issues.

Kubrick's *Dr. Strangelove* poignantly critiques the concept of resilience and adaptation through the absurd proposals of Dr. Strangelove. The irony of Strangelove's words underscores a cynical view of resilience as a solution, highlighting how such advocacy often obscures the underlying social conditions and attempts to present transhistorical theories. These calls for resilience fail to address the fundamental question of why vulnerable communities, who lack the necessary resources, are expected to develop resilience to counteract the negative impacts inflicted by the greed and negligence of more affluent societies.

The discourse surrounding resilience frequently sidesteps accountability, neglecting to address who is responsible, who must pay the price, and who bears the brunt of these impacts. It rarely considers how resilience and adaptation efforts will affect different social classes in disparate ways, often leaving the most vulnerable to suffer while the elite, as depicted in the film, find ways to safeguard themselves.

This selective approach to resilience mirrors the movie's depiction of a society ready to protect its privileged members while ignoring the broader implications for those less fortunate. Kubrick's satire thus serves as a call to critically evaluate the frameworks of resilience and adaptation, urging a deeper consideration of social equity and emancipation. It challenges us to question the systemic inequalities that demand resilience from those least equipped to manage them and to reconsider who should bear the responsibility for mitigating crises.

Such inequalities act as significant deterrents to taking meaningful actions and mobilizing the masses toward climate solutions. They can even set the stage for the rise of right-wing politics. This challenge has long been recognized within the climate movement, which strives to engage workers as beneficiaries of a just transition—whether through a Green New Deal, a Green Industrial Revolution, or similar initiatives. However, waning class consciousness and entrenched systemic structures make this message difficult to convey effectively. In a serious adaptation crisis, for instance, the material benefits associated with whiteness could become even more pronounced, further complicating efforts to galvanize collective action. As Malm and Zetkin Collective (2021, p. 456) note, "*some fractions of the working class would be susceptible to fossil fascism in a mitigation crisis,* as determined as anyone to hold on to what they have got." This susceptibility arises from a desire to maintain existing privileges and resist changes perceived as threatening to their socioeconomic status. The implication is that without addressing these systemic inequalities and reconsidering environmental justice, there is a risk of alienating key demographics and inadvertently fueling reactionary politics. A successful climate movement must therefore first and foremost prioritize emancipation.

The inequalities highlighted in Kubrick's *Dr. Strangelove* extend beyond human communities to encompass the broader ecological web, including different species. This notion of multispecies justice is evident in the film's portrayal of survival strategies that prioritize human life while relegating other species to mere resources. The line, "Greenhouses could maintain plant life. Animals could be bred and *slaughtered,*" satirizes speciesism and underscores the utilitarian view of non-human life forms as expendable commodities in the quest for human survival.

In discussing multispecies justice, Heise (2016) emphasizes the importance of shifting the discourse from fear and apocalypse to a more nuanced understanding of ecological interdependencies, much like Kubrick does through satire. Heise notes that while a shift in discourse might move away from apocalyptic themes, it could also introduce equally troubling narratives of endangerment, such as those related to oppression, slavery, and holocaust. These tropes, rooted in animal welfare and rights movements, suggest that adopting a domestication mindset might reframe rather than resolve the sense of endangerment.

Kubrick's mindful approach to environmental and multispecies justice through satire, rather than fear, challenges us to reconsider the dominant environmentalist discourse in the Chthulucene. By juxtaposing the absurdity of anthropocentric survival plans with the ethical considerations of multispecies justice, the film advocates for a symbiotic understanding of survival and flourishing. Satire invites the audience to rethink prevalent narratives, seeking alternatives that honor the intrinsic value of all species and recognize the interconnectedness of life on Earth.

Nonetheless, Dr. Strangelove's empty promises gain significant traction in the war room, prompting the highest government officials to quickly deliberate on who should be selected to live underground. Strangelove suggests that the selection process "could easily be accomplished with a computer. And a computer could be set and programmed to accept factors from youth, health, sexual fertility, intelligence, and a cross section of necessary skills," a precursor of AI algorithms. Initially, his reasoning may appear logical, yet it falters under closer examination. Absurdity eventually culminates when he assets,

> Of course, it would be absolutely vital that our top government and military men be included to foster and impart the required principles of leadership and tradition. [*Slams down left fist. Right arm rises in stiff Nazi salute.*] Arrrrr! [*Restrains right arm with left.*] Naturally, they would breed prodigiously, eh? There would be much time, and little to do. But ah with the proper breeding techniques and a ratio of say, ten females to each male, I would guess that they could then work their way back to the *present gross national product* [emphasis added] within say, twenty years.

Strangelove envisions political and military leaders breeding extensively, proposing a ratio of ten females to each male, with the expectation that they would restore the gross national product in two decades. This absurd plan underscores the satirical critique of leadership's misguided priorities in crisis scenarios.

The logic of domination governs not only political dominance but also heteropatriarchy. Strangelove's Nazi intervals, which do not receive any particular reaction among the top politicians and military personnel of the county who are actually swayed by his promises, binds physical covetousness and lust for power. Optimized reproduction is supposed to eventually restore the status quo with restoring economic growth. Of course, they ignore the fact that the disaster was originally caused by economic growth in the first place. Their inability to recognize that GDP has failed to secure welfare, justice, peace, security, or sustainability suggests that they are poised to repeat their mistakes, even if afforded a new opportunity. In fact, this is exactly what Major Kong did earlier despite the numerous chances he had to avoid disaster by aborting the false mission.

While the generals indulge in fantasies of extensive sexual encounters with attractive women, the President poses a poignant question to Strangelove: whether survivors might prefer death to enduring unbearable grief and anguish. Strangelove responds, "No sir ... [*Right arm rolls his wheelchair backwards.*] Excuse me. [*Struggles with wayward right arm, ultimately subduing it with a beating from his left.*] Also, when ... when they go down into the mine everyone would still be alive. There would be no shocking memories, and the prevailing emotion will be one of nostalgia for those left behind, combined with a spirit of bold curiosity for the adventure ahead! Ahhhh! [*Right are reflexes into Nazi salute. He pulls it back into his lap and beats it again. Gloved hand attempts to strangle him.*]" Besides prosthetic material agency (Plain, 2023), this scene highlights a profound apathy toward the suffering of others and an indifference to their plight. Kubrick deftly exposes and connects the brutality of war with the slow violence perpetrated by heteropatriarchy, revealing the underlying callousness that governs both political and social structures (see Nixon, 2011).

In the male-dominated war room, the focus shifts from the suffering of others to the enticing prospect of multiple attractive sexual partners in the underground refuge. Remarkably, they have the audacity to present this not as apathy or sheer epicurean selfishness but as a Promethean *sacrifice* for the greater good. Strangelove frames it as, "But it is, you know, a sacrifice required for the future of the human race. I hasten to add that since each man will be required to do prodigious ... *service* along these lines, the women will have to be selected for their sexual characteristics which will have to be of a highly stimulating nature." This self-centeredness and blatant indifference unite DeSadeski, the Soviet ambassador, and the communist-hating Turgidson, as they find common ground in sensual pleasure. Conflicts and disagreements are often set aside when personal interests and physical pleasures are at stake, even at the expense of less fortunate communities or suffering species. However, this coming together does not erase their trivial concerns; even on doomsday, American statesmen remain preoccupied with Soviet expansion, while the Soviet ambassador continues spying and gathering intelligence. By portraying the absurdity of militarization and the potential for anthropogenic annihilation, satire highlights the interconnectedness of environmental issues with broader societal and political dynamics.

## Environmental Concerns

In *Dr. Strangelove or: How I Learned to Stop Worrying and Love the Bomb*, therefore, Stanley Kubrick intricately weaves together societal, political, and technological dynamics through a satirical thread that examines the catastrophic potential of a nuclear confrontation during the Cold War era. While environmental issues are not overtly central to the film, they are subtly interwoven with the main themes, highlighting their interconnectedness. Set during the height of Cold War tensions, the narrative unfolds as a US general, driven by personal insecurities, initiates a nuclear attack on the USSR, prompting a frantic scramble among military and political officials to prevent Armageddon. This catastrophe is rooted in a man's perceived failure to conform to heteropatriarchal and hegemonic masculine norms and echoes the theme of gender power dynamics reminiscent of Aristophanes' *Lysistrata*.

The film underscores several key themes that link environmental concerns with political and social structures. The absurd and reckless behavior of the political and military elite reflects a broader critique of how political hubris can lead to environmental neglect. One can argue that the same lack of foresight and accountability that precipitates political conflicts and wars can also result in environmental degradation. Dr. Strangelove, with his history as a former Nazi scientist, symbolizes the perilous overreach of technology when wielded by ideologically driven leaders. This character echoes contemporary concerns about the environmental impact of unchecked technological advancements and the rise of right-wing climate denialism alongside increasing securitization and militarization like the world has been experiencing in the second decade of the twenty-first century.

The film's narrative, which revolves around international conflict and cooperation, underscores the global nature of environmental issues. It illustrates how crises like nuclear fallout transcend national borders, highlighting the necessity for international collaboration and comprehensive policies to address global environmental challenges. This allegorical reading extends the film's critique beyond its immediate focus on nuclear warfare to suggest broader implications for sustainability.

By examining the absurdities of political and military decision-making, the movie invites viewers to consider the broader societal and environmental impacts of these actions. In doing so, it challenges audiences to reflect on the responsibilities of leaders and societies in preventing environmental degradation and fostering a more sustainable and emancipating future.

Through its depiction of the absurdity of militarization, *Dr. Strangelove* indirectly underscores the interconnectedness of environmental issues with broader societal and political dynamics. While the film does not directly address environmental urgencies such as climate change, it touches on related issues, including the environmental impact of nuclear weapons. The narrative highlights the dangers of nuclear militarism as a significant environmental concern.

*Dr. Strangelove* uses the backdrop of nuclear warfare to delve into how the interplay between political, military, and technological forces can lead to profound environmental consequences. Through its dark satire and critical portrayal of Cold War politics, the film can indirectly call for heightened environmental consciousness and responsibility within the broader societal and political framework. In doing so, it reflects on two major themes: nuclear catastrophe as environmental disaster and Mutually Assured Destruction and environmental ethics.

The film's principal premise, centered on the potential for a nuclear holocaust, serves as a stark premonition of an extreme environmental disaster. The imminent doomsday scenario depicted in the movie underscores the vulnerability of the environment to political decisions and technological advancements. The narrative invites reflection on how human actions can precipitate catastrophic environmental outcomes, prompting a reevaluation of the strategies employed to manage global challenges.

The portrayal of the doctrine of mutually assured destruction (MAD) poses a profound ethical dilemma. Reliance on the threat of total annihilation as a means of maintaining peace raises critical questions about the ethical implications of such a strategy (Lindley, 2001; Roberts, 2021). This doctrine not only reflects the precarious nature of Cold War politics but also highlights broader ethical issues regarding the planet. It underscores the moral responsibility of leaders to protect the environment, rather than endanger it, by prioritizing sustainable and peaceful coexistence over destructive power dynamics.

The stakes are elevated not merely because bombing the Soviet Union risks initiating a nuclear war, but because the USSR has constructed a "doomsday machine." This "ultimate weapon" is not designed to target conventional military threats but to obliterate all life on the planet in an automated act of omnicide. The film reveals that when detonated, the device will unleash enough lethal radioactive fallout to render the planet's surface "as dead as the moon" within ten months. The detonation creates a "doomsday shroud," a lethal radioactive cloud that will encircle the Earth for ninety-three years, ensuring complete and indiscriminate destruction.

This doomsday machine is not a traditional weapon of defense; it is engineered to self-trigger and is irreversible. As the narrative unfolds, it is revealed that "it is designed to explode if any attempt is ever made to untrigger it." This characteristic underscores its terrifying nature, as it cannot be recalled or deactivated once set in motion. Unlike conventional defense systems that aim to repel enemies while sparing the defender, the doomsday device is designed to obliterate all life forms indiscriminately, eradicating not just human beings but all creatures on the planet.

The film's depiction of the doomsday machine serves as a chilling metaphor for the ultimate consequences of unchecked technological advancements and the perilous allure of absolute power. It highlights the folly of creating weapons that, rather than ensuring security, guarantee mutual destruction. In doing so, *Dr. Strangelove* critiques the logic of deterrence that dominated Cold War politics, urging a reconsideration of the ethical responsibilities associated with wielding such destructive capabilities (Kirshner, 2001; Maloney, 2020). In a fresh reading, the doomsday scenario is eerily similar to the total destruction that climate change will bring on all critters living on the planet. Extinction, habitat loss, and extreme weather events will make the planet uninhabitable for all species.

In many ways, there are striking parallels between the doctrine of MAD and our current environmental trajectory. MAD, as a military doctrine, relies on the principle that two or more opposing sides have enough nuclear capability to destroy each other, thus deterring any actual use of nuclear weapons. The threat of total annihilation is meant to maintain peace by making the cost of conflict unacceptably high. However, the absurdity of this doctrine is glaringly apparent, and some within the government have attempted to resist it, though their efforts have ultimately failed.

The reasons for this failure can be traced back to capitalism and consumerism. The pressures of capitalism are evident in the acknowledgment that "there are those of us who fought against it, but in the end, we could not keep up with the expense involved in the arms race, the space race, and the peace race." Here, the economic demands of maintaining competitive military and technological advancements override ethical considerations. Consumerism further complicates matters, as forged societal desires drive decision-making: "And at the same time our people grumbled for more nylons and washing machines." These consumer demands dictate market economics, making it seemingly cheaper to risk annihilation than to curtail consumption and competition. This is underscored by the rationale, "Our doomsday scheme cost us just a small fraction of what we'd been spending on defense in a single year. But the deciding factor was when we learned that your country was working along similar lines, and we were afraid of a doomsday gap."

The movie reveals that the US had also considered constructing a doomsday machine, unbeknownst to the President. This revelation highlights the disconnect between leadership and the actions taken in the name of national security, reflecting a broader critique of how economic and military imperatives can lead to decisions that endanger all life on Earth. The parallel to our ecological trajectory suggests that without significant changes in priorities, humanity risks replicating the MAD doctrine's folly on a planetary scale.

Constructing the doomsday device is one matter, but designing it to self-trigger and be irreversible—akin to the challenges posed by climate change beyond the tipping points—is another. When asked about the impossibility of deactivating the device, Dr. Strangelove explains, "Mr. President, it is not only possible, it is essential. That is the whole idea of this machine, you know. Deterrence is the art of producing in the mind of the enemy … the *fear* [emphasis added] to attack. And so, because of the automated and irrevocable decision-making process which rules out human meddling, the doomsday machine is terrifying. It's simple to understand. And completely credible, and convincing."

His explanation exposes the irrationality of relying on fear as a deterrent, intended to prevent war by instilling terror in potential aggressors (see Cooke, 2007). The expectation was that fear would establish peace. However, the film demonstrates in real time the failure of this fear-based approach. This is analogous to the use of fear in environmental communication and narratives, which can also falter. Kubrick employs satire instead of somber fear to advocate for change, suggesting that alternative strategies may be more effective in prompting action and addressing global challenges.

While Kubrick critiques capitalism and consumerism, he remains vigilant not to reinforce simplistic binaries between political ideologies. Although it is the communist USSR that constructs the doomsday weapon, it is a US officer, known for his anti-communist stance, who triggers it. This nuance highlights that the problem lies not within any single ideology, but in the broader dynamics of militarism and power (Ackerman, 2015). Kubrick's approach reveals that the potential for catastrophe transcends ideological boundaries, critiquing the global arms race and the flawed logic underpinning Cold War strategies. Rather than reinforcing binaries, the film exposes the universal follies of militarism and the existential threats posed by unchecked technological and political ambitions, urging a reexamination of how different ideologies contribute to these dynamics.

The film opens with the grave voice-over narration: "In order to guard against surprise nuclear attack, America's Strategic Air Command maintains a large force of B52 bombers, airborne 24 hours a day. Each B52 can deliver a nuclear bomb load of 50 megatons, equal to 16 times the total explosive force of all the bombs and shells used by all armies in World War II. Based in America, the Airborne Alert Force is deployed from the Persian Gulf to the Arctic Ocean. But they have one geographical factor in common. They are all two hours from their target inside Russia." The gravitas of the voice over narration betrays the otherwise comedic nature of the movie (Sperb, 2006).

When the crew onboard a B52 receives the top-secret attack code, they are caught in mundane activities: one officer is playing cards, another is looking at a nude picture, and a third is reading a book. Initially, they can't believe what they see: "Ain't nobody ever got the go code yet. And old Ripper wouldn't be giving us plan R unless them Russkies had already clobbered Washington and a lot of other towns with a sneak attack."

Plan R, as explained to President Muffley in the War Room, is a contingency measure:

> Plan R is an emergency war plan in which a lower echelon commander may order nuclear retaliation after a sneak attack if the normal chain of command is disrupted. You approved it, sir. You must remember. Surely you must recall, sir, when Senator Buford made that big hassle about our deterrent

> lacking credibility. The idea was for plan R to be a sort of retaliatory safeguard. ... I admit the human element seems to have failed us here. But the idea was to discourage the Russkies from any hope that they could knock out Washington, and yourself, sir, as part of a general sneak attack, and escape retaliation because of lack of proper command and control. ... As you may recall, sir, one of the provisions of plan R provides that once the go code is received the normal SSB radios in the aircraft are switched into a special coded device, which I believe is designated as CRM114. Now, in order to prevent the enemy from issuing fake or confusing orders, CRM114 is designed not to receive at all, unless the message is preceded by the correct three letter code group prefix.

The understatement in acknowledging the "failure of the human element" amid the potential for indiscriminate annihilation heightens the satirical effect. Moreover, the logic behind the irreversibility of the plan similar to that of the doomsday device further highlights the absurdity and peril of systems designed to operate autonomously.

After the base confirms the attack message, the commander of the B52 swaps his flight helmet for a Texas cowboy hat and delivers an emotional speech to rally his crew for nuclear combat, appealing to their nationalist sentiments. He assures them, "This thing turns out to be half as important as I figure it just might be, I'd say that you're all in line for some important promotions and personal citations when this thing's over with. That goes for every last one of you, regardless of your race, color, or your creed. Now, let's get this thing on the hump. We got some flying to do." In saying this, the patriotic officer inadvertently nods toward rampant racism and discrimination in the US society.

The film then cuts to General Turgidson's half-naked secretary, clad in a bikini and lounging in his room, highlighting Turgidson's lack of urgency or alarm upon hearing the news of the nuclear attack. This starkly contrasts with the deranged General Ripper, whose hyper-nationalism drives him to initiate a nuclear war. Ripper's paranoia is evident as he declares, "I can no longer sit back and allow Communist infiltration, Communist indoctrination, Communist subversion, and the international Communist conspiracy to sap and impurify all of our precious bodily fluids."

Having issued the go code, Ripper contacts Strategic Air Command headquarters, urging full commitment from the Administration since the attack cannot be recalled. He concludes with, "God willing, we will prevail in peace and freedom from fear and in true health through the purity and essence of our natural fluids. God bless you all." The absurdity of his final statement leaves the Administration perplexed over what the meaning of that last phrase might be. This underscores the irrationality and dangerous extremism that drive the movie's characters, reflecting Kubrick's critique of the absurdities inherent in Cold War mentalities and the peril of unchecked nationalism (see also Higgins, 2017; Maloney, 2020).

General Ripper's idiosyncrasies extend beyond mere patriotism or anti-communist fervor; his actions are ultimately driven by personal insecurities in a heteropatriarchal society. His fixation on the purity of "precious bodily fluids" emerges as a bizarre rationale for his actions, veering into conspiracy. Ripper tells Mandrake, "Water is the source of all life. Seven tenths of this earth's surface is water. Why, you realize that … seventy percent of you is water. And as human beings, you and I need fresh, pure water to replenish our precious bodily fluids … Have you ever heard of a thing called fluoridation? Fluoridation of water? Do you realize that fluoridation is the most monstrously conceived and dangerous Communist plot we have ever had to face?"

Ripper further elaborates, "Mandrake, do you realize that in addition to fluoridated water, why, there are studies underway to fluoridate salt, flour, fruit juices, soup, sugar, milk, ice cream? Ice cream, Mandrake. Children's ice cream? … Fluoridation first began … [in] Nineteen hundred and forty-six. Nineteen forty-six, Mandrake. How does that coincide with your postwar commie conspiracy, huh? It's incredibly obvious, isn't it? A foreign substance is introduced into our precious bodily fluids without the knowledge of the individual, and certainly without any choice. That's the way your hard core commie works."

These conspiracy theories eerily parallel those that gained traction during the pandemic and in the rise of extreme right-wing governments globally. Figures like Robert F. Kennedy Jr., United States Secretary of Health and Human Services, espousing similar baseless beliefs, now hold official positions of power in the US Administration, revealing the enduring appeal of such ideas. Despite the six decades separating Kennedy's era from Ripper's, the resonance between their ideas is startling.

This narrative unfolds against the backdrop of genuine and serious concerns, where the action taken—initiating a nuclear attack—exacerbates rather than addresses these issues. Kubrick's film underscores the folly of using destructive means to tackle urgent challenges, highlighting the irrationality and danger of conflating personal or partisan paranoia with global policy decisions.

*Dr. Strangelove* is a richly layered film that goes beyond merely critiquing Cold War militarism and the environmental hazards of nuclear weapons. Kubrick also takes aim at heteropatriarchy, highlighting how it entails a form of slow violence that can eventually erupt into full-scale destruction. When asked about the origins of his conspiracy theories, Ripper responds, "I first became aware of it, Mandrake, during the physical act of love ... Yes, a profound sense of fatigue, a feeling of emptiness followed. Luckily, I was able to interpret these feelings correctly: loss of essence ... I can assure you it has not recurred, Mandrake. Women ... women sense my power, and they seek the life essence. I do not avoid women, Mandrake, but I do deny them my essence." This dialogue underscores the toxic demands of heteropatriarchy and the fetishization of masculinity (Kolker, 2017; Linden, 1977; Maloney, 2020), issues that continue to manifest in contemporary society through phenomena like the promotion of "alpha males" and practices such as "testicle tanning" in the US. These societal pressures lead to derangements that can have far-reaching consequences.

In addition to heteropatriarchy, religious zeal also plays a role in Ripper's irrational actions. His derangement is not solely the result of conspiracy theories. He later reveals, "I happen to believe in a life after this one, and I know I'll have to answer for what I've done. And I think I can," just before committing suicide. This statement highlights how irrational religious fervor can also motivate absurd and dangerous behavior, a theme that remains relevant as religious zeal resurfaces in contemporary societies.

Kubrick's film thus serves as a multifaceted critique, addressing not only the overt dangers of militarism and nuclear proliferation during the Cold War but also the underlying social and psychological factors that drive individuals toward destructive actions. By exposing these dynamics, *Dr. Strangelove* invites viewers to reflect on the complex interplay of ideologies that shape global conflicts and urgencies.

Striving to correct "an imperfect world" (Tonn, 2017, p. 246), Stanley Kubrick artfully critiques the convergence of various destructive forces—economic, political, and military rivalries; consumerism and capitalism; heteropatriarchy; and religious zealotry—all culminating in a nuclear catastrophe threatening total annihilation. Through its dark satire (Eaton, 2012; Lemos, 2012; Shaw & Youngblood, 2017), the film underscores the reckless pursuit of power at the expense of rationality, serving as a cautionary reminder of the dangers of prioritizing short-term interests over long-term well-being and sustainability. As McDonald (2022, p. 27) contends "the explosion of atomic laughter obliterates humanist discourse to create a newly flattened philosophical terrain no longer riveted by the strictures of Enlightenment thought." In other words, Kubrick employs satire to flatten the philosophical terrain and open a space for the rise of posthuman interspecies relationships. Elsewhere, she contends, the movie features "affirmative laughter that does the important philosophical work of imagining an affectively charged 'outside' to the genocidal trajectory of humanist thought" (McDonald, 2020, pp. 282–283).

The characters, epitomized by figures like General Jack D. Ripper and Dr. Strangelove, highlight the absurdity of political and military decision-making processes driven by irrational fears, unrealistic social normative demands, and ideological obsessions. This critique of decision-making is mirrored in contemporary environmental issues where short-term economic and political gains frequently overshadow sustainable practices, leading to environmental degradation and resource depletion.

Moreover, the reckless handling of nuclear weapons in the film serves as a metaphor for the broader issue of irresponsible treatment of the planet. Just as the uncontrolled proliferation and potential use of nuclear weapons threaten global security, unsustainable environmental practices pose existential threats to the planet's health and future generations. The concluding Armageddon highlights the dire consequences of neglecting sustainability, paralleling the threat of nuclear apocalypse. It starkly warns that ignoring long-term impacts, whether in nuclear policy or environmental management, can lead to irreversible damage and catastrophe.

Kubrick's satire also calls for a heightened sense of ethical responsibility among leaders, suggesting that the well-being of the ecozone must be prioritized over immediate power gains. By showcasing the ultimate failure of political and military leaders to prevent disaster, *Dr. Strangelove* advocates for policies that ensure the sustainable use of resources. The film underscores the interconnectedness of major global issues, like nuclear security and environmental sustainability, implying that effective solutions require international cooperation and a commitment to collective long-term goals rather than isolated, short-term interests.

*Dr. Strangelove* offers a potent critique of the reckless pursuit of power and its detrimental effects on rational policy-making and long-term sustainability. It advocates balancing short-term interests with the imperative of protecting and sustaining both nuclear arsenals and the planet for future generations. Kubrick's work thus remains a powerful and relevant commentary on the ethical and practical considerations necessary to avert global catastrophe and secure a sustainable future.

*Dr. Strangelove* further foregrounds missed opportunities and the relentless march toward catastrophe as in climate and environmental action. Throughout the film, the characters are presented with numerous chances to avert disaster: attempts to recall the bombers are thwarted by malfunctioning radios, Russian defenses manage to down three bombers, and the remaining plane faces significant challenges like damage, fuel leakage, and jammed bomb bay doors. Despite these numerous second and third chances, the characters remain adamant in triggering the doomsday machine. This mirrors the global procrastination and akrasia in addressing climate change, where each delay risks setting off irreversible chain reactions.

An absurd sense of idealized patriotism drives the characters, celebrated in the film by the iconic image of the Yankee riding the bomb to his and everyone's doom. Similarly, the fear of economic consequences and drastic lifestyle changes often postpones climate action, ignoring the slowly unfolding but inevitable consequences. In one scene, Mandrake's urgent need to relay the recall code to the President leads him to threaten Guano into shooting a vending machine for change. Guano's retort, "Ok. I'm gonna get your money for you. But if you don't get the President of the United States on that phone, you know what's going to happen to you?" with the punchline, "You're going to have to answer to the Coca-Cola Company," highlights the immense power of corporations, a subtle nod to their influence over political and environmental decision-making.

The movie ends with Strangelove shouting "…sir! I have a plan. Heh," while he manages to stand up out of his wheelchair in disbelief. He pauses, realizing that he is standing, cries in disbelief, "Mein Führer, I can walk!" The renewed rise of fascism is immediately followed by multiple scenes of exploding bombs unfold to the tune of "We'll Meet Again."

## References

Ackerman, A. (2015). A "fluid" truism: *Dr. Strangelove's* confrontation of the malleability of knowledge and truth. *Young Scholars in Writing, 12*, 75–82. https://youngscholarsinwriting.org/index.php/ysiw/article/view/211

Barca, S. (2024). *Workers of the earth: Labour, ecology and reproduction in the age of climate change*. Pluto Press.

Broderick, M. (1993). Surviving armageddon: Beyond the imagination of disaster. *Science Fiction Studies, 20*(3), 362–382. http://www.jstor.org/stable/4240277.

Broderick, M. (2017). *Reconstructing strangelove: Inside Stanley Kubrick's nightmare comedy*. Columbia University Press. https://doi.org/10.7312/brod17708

Buck, H. J. (2019). *After geoengineering: Climate tragedy, repair, and restoration*. Verso Books.

Caradonna, J. L., et al. (2015, May 6). *A degrowth response to An ecomodernist manifesto*. Resilience.org. https://www.resilience.org/stories/2015-05-06/a-degrowth-response-to-an-ecomodernist-manifesto/

Carton, W., & Malm, A. (2025). *The long heat: Climate politics when it's too late*. Verso.

Chaudhary, A. S. (2024). *The exhausted of the Earth: Politics in a burning world*. Penguin.

Cooke, E. F. (2007). Understanding the enemy: The dialogue of fear in *Fear and Desire* and *Dr. Strangelove*. In J. J. Abrams (Ed.), *The philosophy of Stanley Kubrick* (pp. 9–31). The University Press of Kentucky.

Crist, E. (2016). The reaches of freedom: A response to an ecomodernist manifesto. *Environmental Humanities, 7*(1), 245–254. https://doi.org/10.1215/22011919-3616452

Deutsch, J. I. (2016). 'No communist could dream of a more effective anti-American film': Dr. Strangelove and its red reviewers. *Historical Journal of Film, Radio and Television, 36*(1), 40–52. https://doi.org/10.1080/01439685.2015.1134108

*Dr. Strangelove or: How I Learned to Stop Worrying and Love the Bomb*. (1964). [Film]. Directed by S. Kubrick. Columbia Pictures. https://archive.org/details/dr.-strangelove-1964

Eaton, M. (2012). Dark comedy from *Dr. Strangelove* to the dude. In A. Horton & J. E. Rapf (Eds.), *A companion to film comedy* (pp. 315–339). Wiley. https://doi.org/10.1002/9781118327821.ch15

García Mainar, L. M. (1994). Kubrick's *Dr. Strangelove*: Desire and the fragmenting of character. *Revista Alicantina de Estudios Ingleses, 7*, 67. https://doi.org/10.14198/raei.1994.7.06

Haraway, D. J. (1985). A cyborg manifesto: Science, technology, and socialist-feminism in the late twentieth century. *Socialist Review, 80*, 65–108.

Hardin, G. (1968). The tragedy of the commons. *Science, 162*(3859), 1243–1248. https://doi.org/10.1126/science.162.3859.1243

Heise, U. K. (2016). *Imagining extinction: The cultural meanings of endangered species*. The University of Chicago Press.

Heron, K. (2022, September 7). The great unfettering. *New Left Review*. https://newleftreview.org/sidecar/posts/the-great-unfettering

Higgins, S. (2017). Purity of essence in the cold war: *Dr. Strangelove*, paranoia, and bodily boundaries. *Textual Practice, 32*(5), 799–820. https://doi.org/10.1080/0950236x.2016.1277549

Huber. (2022, September 7). Mish-mash ecologism. *New Left Review*. https://newleftreview.org/sidecar/posts/mish-mash-ecologism

Kirshner, J. (2001). Subverting the cold war in the 1960s: Dr. Strangelove, the manchurian candidate, and the planet of the apes. *Film & History: An Interdisciplinary Journal of Film and Television Studies, 31*(2), 40–44. https://doi.org/10.1353/flm.2001.a400700

Kolker, R. P. (2017). *The extraordinary image: Orson Welles, Alfred Hitchcock, Stanley Kubrick, and the reimagining of cinema*. Rutgers University Press. http://www.jstor.org/stable/j.ctt1ht4v39

Latour, B. (2015, June). Fifty shades of green: Presentation to the panel on modernism at the Breakthrough Dialog, Sausalito. *Bruno Latour Official Website*. http://www.bruno-latour.fr/sites/default/files/downloads/00-BREAKTHROUGH-06-15_0.pdf

Latour, B. (2017). *Facing Gaia: Eight lectures on the new climatic regime*. Polity.

Lemos, M. (2012). Dr. Jeems Joker or: How I learned to stop worrying and love *Finnegans Wake*. In M. A. Alonso (Ed.), *Weaving new perspectives together: Some reflections on literary studies* (pp. 74–93). Cambridge Scholars Publishing.

Linden, G. W. (1977). *Dr. Strangelove* and erotic displacement. *Journal of Aesthetic Education, 11*(1), 63–83. https://doi.org/10.2307/3331863

Lindley, D. (2001). What I learned since I stopped worrying and studied the movie: A teaching guide to Stanley Kubrick's *Dr. Strangelove*. *PS: Political Science and Politics, 34*(3), 663–667. http://www.jstor.org/stable/1353558

Maguire, L. (2018). The scientific world faced with the apocalypse of its own creation: A comparison of *Dr. Strangelove* and *Nine Days of One Year*. *The Journal of Popular Culture, 51*(4), 956–971. https://doi.org/10.1111/jpcu.12704

Maland, C. (1979). *Dr. Strangelove* (1964): Nightmare comedy and the ideology of liberal consensus. *American Quarterly, 31*(5), 697–717. https://doi.org/10.2307/2712432

Malm, A., & Carton, W. (2024). *Overshoot: How the world surrendered to climate breakdown*. Verso.

Malm, A., & Zetkin Collective. (2021). *White skin, black fuel: On the danger of fossil fascism*. Verso.

Maloney, S. M. (2020). *Deconstructing Dr. Strangelove: The secret history of nuclear war films*. Project MUSE. https://muse.jhu.edu/book/74964

Malthus, T. R. (1798). *An essay on the principle of population: As it affects the future improvement of society, with remarks on the speculations of Mr. Godwin, M. Condorcet, and other writers*. J. Johnson. Retrieved from http://www.esp.org/books/malthus/population/malthus.pdf

McDonald, F. (2020). The last laughs of doomsday humor. In *Apocalypse in American literature and culture* (pp. 281–292). Cambridge University Press. https://doi.org/10.1017/9781108663557.022

McDonald, F. (2022). *Posthumorism: The modernist affect of laughter.* Bloomsbury Academic.

Nixon, R. (2011). *Slow violence and the environmentalism of the poor.* Harvard University Press. https://doi.org/10.2307/j.ctt2jbsgw

Plain, G. (2023). *Prosthetic agency: Literature, culture, and masculinity after World War II.* Cambridge University Press.

Robbins, P., & Moore, S. A. (2015, June 19). Love your symptoms: A sympathetic diagnosis of An ecomodernist manifesto. *Entitle Blog.* https://entitleblogdotorg3.wordpress.com/2015/06/19/love-your-symptoms-a-sympathetic-diagnosis-of-the-ecomodernist-manifesto/

Roberts, B. (2021). On 'Campaigning' for nuclear deterrence. In S. Stephan Frühling & A. O'Neil (Eds.), *Alliances, nuclear weapons and escalation: Managing deterrence in the 21st century* (1st ed., pp. 187–198). ANU Press. http://www.jstor.org/stable/j.ctv25m8dp0.21

Saito, K. (2024). *Slow down: The degrowth manifesto.* Astra Publishing House.

Shaw, T., & Youngblood, D. J. (2017). Cold war sport, film, and propaganda: A comparative analysis of the superpowers. *Journal of Cold War Studies, 19*(1), 160–192. https://www.jstor.org/stable/26925679

Simon, W. (2003). *Dr. Strangelove* or: The apparatus of nuclear warfare. In R. Allen & M. Turvey (Eds.), *Camera obscura, camera lucida: Essays in honor of Annette Michelson* (pp. 215–230). Amsterdam University Press. https://doi.org/10.1515/9789048505067-013

Sperb, J. (2006). *The Kubrick facade: Faces and voices in the films of Stanley Kubrick.*

Sperrin, D. (2025). *State of ridicule: A history of satire in English literature.* Princeton University Press. https://doi.org/10.2307/jj.22679818

Stillman, G. B. (2008). Two of the MADdest scientists: Where Strangelove Meets Dr. No; or, unexpected roots for Kubrick's Cold war classic. *Film History, 20*(4), 487–500. http://www.jstor.org/stable/27670748

Tonn, M. B. (2017). Making and unmaking political mischief: Trickster influences in the rhetorical humor of the 1960s. In R. J. Jensen (Ed.), *Social controversy and public address in the 1960s and early 1970s: A rhetorical history of the United States* (Vol. 9, pp. 221–250). Michigan State University Press. https://doi.org/10.14321/j.ctt1vjqqkq.11

Traub, C. (2018). From the grotesque to nuclear-age precedents: The modes and meanings of cli-fi humor. *Studies in the Novel, 50*(1), 86–107. https://www.jstor.org/stable/48559258

CHAPTER 3

# Navigating Environmental Satire in the Culture Industry

Some researchers and activists are not convinced about treating the urgency of climate change as a serious issue within a satiric framework as it can potentially undermine its gravity (Little, 2022), but this is not the only reservation about the efficacy of environmental satire. Constraints of the regime of communication under capitalism is another serious concern. *Don't Look Up* (McKay, 2021) is a good example to illustrate the paradoxes of environmental satire enmeshed in the culture industry (Horkheimer & Adorno, 2016). The movie decries politicians and the media and censures science communication in hopes of advocating decisive and timely action on climate change. It acknowledges the power, resources, and impacts of governments and corporations in addressing climate crises are not comparable to those of individuals, and underlines the disparity in the responsibilities of individuals as well as institutions. *Don't Look Up* also criticizes the role of academics and the media in addressing climate change and insists on the intersectionality of environmental crises with adjacent socio-political issues including class and race.

Directed by Adam McKay, *Don't Look Up* was produced and distributed by Netflix in 2021. The movie was produced with a budget of US$75 million and grossed almost US$800 million. It tells the story of a PhD candidate who discovers a comet which turns out to be on a direct collision course with the Earth. The plot unfolds with her and her supervisor's desperate attempts to warn politicians and the public about the impending



M. Zekavat, *Leveraging Satire for Environmental Advocacy*, https://doi.org/10.1007/978-3-032-04838-7_3

disaster of extinction, but they eventually fail to persuade and galvanize the government, media, and corporations to take immediate action in deflecting the asteroid and saving the planet. The disastrous impact of the comet is a thin satirical allegory for the annihilation that our inaction on climate change will bring and the response of different stakeholders to the urgency of the crisis. The movie assumes the form of a satirical commentary, critiquing the apathy of governmental bodies, political entities, prominent individuals, and the media towards the pressing concerns posed by the climate crisis. It frames the gradual and remote risk of unfolding climate change as the immediate danger of an asteroid colliding with the planet to underscore the urgency of the crisis.

*Don't Look Up* opens with Kate Dibiasky (Jennifer Lawrence), a PhD candidate at Michigan State University, observing the sky. She is listening to Wu-Tang Clan's "Wu-Tang Clan Ain't Nuthing Ta F' Wit" (2013) and singing along: "inside the room/Dr. Doom, prepare for the boom." The opening lyric not only foreshadows the eventual disastrous outcome but also locates the movie in the political framework of racial and social justice even before the opening credits roll.

Next, we see Kate, Dr. Randall Mindy (Leonardo DiCaprio), her supervisor, and their team celebrate their discovery shortly before realizing that the comet is headed toward the Earth. The screen fades black and a joke from Jack Handey, an American humorist, appears on the screen: "I want to die peacefully in my sleep like my grandfather, not screaming in terror like his passengers." This is yet another hilarious clue to the disastrous denouement.

The movie largely explores the reasons behind climate change inaction and accuses institutions including political bodies, academic institutions, media, corporations, and religion for the inertia, while underscoring intersectionality and environmental justice.

## Political Bodies

On discovering the danger that the hurtling comet poses for the planet, Randall, calls Dr. Jocelyn Calder (Hettienne Park), the Administrator of NASA at Kennedy Space Center, to warn her about the impending disaster, but she is more concerned with breaking the news of losing US$8 billion in funding to her team. She eventually connects them with Dr. Teddy Oglethorpe (Rob Morgan), the head of the Planetary Defense Coordination Office. Much to Randall and Kate's amazement, this is the name of a real office. The viewer is reassured about the verisimilitude by these words appearing in large font on the screen: "The Planetary Defense Coordination Office is a real place," followed by a display of their logo. The irony is that we are prepared to defend the planet against asteroids but are not willing to stop the havoc we have been wreaking on the planet, although the eventual impact will be the same, i.e., extinction of all forms of life on Earth.

This is also a reminder of the governmental bodies that are already in place for addressing climate change in the US include The Office of Climate Change, US Department of State; United States Environmental Protection Agency (EPA); National Climate Task Force, The White House; and the plethora of centers, think tanks, and academic institutions. In drawing attention to the costly and convoluted organizational labyrinth, the movie criticizes the sheer inefficacy of these structures. All estimations indicate that the comet will directly hit the Earth in 6 months 14 days in "an extinction-level event," yet the array of governmental bodies eventually fails to take effective and timely action despite the extravagant funding they receive. They all have failed to produce the desired effect so far despite the conspicuous urgency of the crisis.

Teddy, who represents a socially responsible scientist with intact academic integrity, takes immediate action by asking Jocelyn to immediately arrange for Kate and Randall to go to Washington, DC, where they are admitted to the White House and meet Teddy and General Themes who is supposed to oversee the developments for the Pentagon. The group is scheduled to meet the President who "is famously late." Politicians are again criticized not only for their mismanagement of priorities but also for their failure to take timely action on climate change.

Madam President Janie Orlean (Meryl Streep) eventually makes her entrance. Complaining that "you told me he was clean," she ignores Teddy and her bodyguards push him out of the way. Teddy calls out Jason Orlean (Jonah Hill)—Chief of Staff and Presidents' son—saying, "Come on, man. What the hell? This is real world." He is reassured that the President needs just five minutes as they face a problem with their Supreme Court nominee. Nepotism is a clear target of satire as key decision making positions and influential people are not assigned because of their merits. Moreover, the Administration trivializes a catastrophic event that will terminate life on the planet while prioritizing a trifling because it has consequences for their political future. We hear the White House staff singing happy birthday and loud laughter from the Oval Office. Politicians' indifference and irresponsibility is further underlined when we hear a woman in the background saying, "Just take a message, and then, stuff like this happens." Political and economic factors determine collective environmental decisions and actions as politicians worry about the perceived political risks of environmental policies. Their goals and aspirations are in conflict with sustainable causes and the wellbeing of all species. Such conflicting goals and aspirations can further cause cognitive dissonance in the decision-making process where the perceived costs of climate action outweigh its economic and political promises.

While they are waiting, General Themes brings them bottled water and snacks, saying, "They charge an arm and a leg for this stuff. Ten apiece ought to do it." The three-star general who works at Pentagon exuberantly charges them for the small snacks. Teddy's reaction implies that he is aware that they do not really have to pay anything. In return, Themes does not even return Teddy's change. After a while Themes receives a message and tells everyone that he is supposed to go to Okinawa to "quell the natives" to which Teddy replies, "Tell your daughter I said hello." Teddy again implies that the prevaricating Themes is not leaving on any official business, but rather this is a frivolous excuse to bail out of the situation. In the next scene, we see that Kate goes and takes more snacks and asks a woman who is sitting there where she should pay for them. The woman responds, "It's free. … It's the White House." The incident continues to haunt Kate for the rest of the plot. She eventually persuades herself to interpret Themes' behavior as power play and a way to assert his military power over their scientific knowledge. The irony is that petty people concerned with petty games are assigned to be in charge of existentially consequential crises.

Kate, Randall, and Teddy are eventually informed by the Chief of Staff, "Hey, peeps, uh, it's not looking like it's gonna happen tonight." When Randall asks him if the President knows why they are there, Jason only slams the door in his face. After a long and urgent flight in a military plane and 7 hours of wait at the White House, they are lodged in a motel, having made no progress despite the urgency of the situation. Later, Kate confides the story to her boyfriend, a journalist, and learns that the President is facing backlash for nominating Sheriff Wade Conlon for the Supreme Court. Conlon is apparently not qualified to serve on the Supreme Court as he holds no law degree; moreover, he is infamous for his record of "shoot first and ask questions later" which is the last quality one might seek in a Justice. Besides, new revelations have brought his past as a nude life model into light. Later, we learn that he is also identified as appearing in a porn cable show. His reaction to all these scandals is: "I have no shame in what I did." One might wonder why the President insists on nominating him for the post, but the reason has to wait for a while.

Kate, Randall, and Teddy are called in to meet the President the following day. Randall is petrified with anxiety; he cannot breathe well. "I heard there is an asteroid, or a comet, or something that you don't like the looks of," says the President of the US. Randall is confronted with sheer ignorance and mockery on the part of the President and her son/Chief of Staff when he explains the gravity of the situation. Eventually, Teddy has to interrupt the conversation to drive the message home by dumbing it down to, "Madam President, this comet is what we call a planet-killer." After the scientists finally manage to overcome ignorance on the part of politicians, the second intention-action impediment kicks in when the President asks, "So how certain is this?" Uncertainty is one of the psychological barriers that can hinder meaningful pro-environmental behavior that is also frequently manipulated at socio-political level. Any sign of uncertainty about climate change might be perceived as reason enough to postpone action, or sheer lack of action. Politicians and pundits frequently resort to this by saying that we should wait until science is in. This poses a conundrum for scientists: on the one hand, they cannot claim 100% certainty so as to abide by scientific methodology; on the other hand, a slightest expression of legitimate scientific margin of error or uncertainty is abused to hinder pro-environmental behavior at both individual and collective levels. Seeking reconciliation, a staff member attending the meeting suggests, "Can we call it a potentially significant event?" Besides foregrounding ignorance and uncertainty as two barriers hindering climate action, the movie also chastises how politicians manipulate discourse in order to frame the situation in a light that is favorable to them.

While attendees are debating the certainty of the collision, the President says, "Call it 70% and let's just move on," as if they are bargaining. When Kate objects to this prevarication, the President insists, "you cannot go around saying to people that there's a 100% chance that they're gonna die. ... It's just nuts." An opportunist, she clearly prioritizes her own political future and that of her party over the survival of life on the planet. It is not even worth trying if saving the planet means she and her party will lose office. She further stokes uncertainty when she suggests that the White House should get some of their scientists on this which provokes Teddy to list their credentials. Jason laughs at the words "Michigan State," adding, "Come on bro!"

The President eventually intervenes by saying, "Let's bottom-line this. What is this gonna cost *me*? You know, what's the ask here?" The deal-maker is still bargaining as if the consequences of an event that threatens the existence of life on the planet is negotiable. The movie criticizes how climate change is subject to political interests and polarization in the US whereas a planetary threat should be perceived far from partisan. It therefore manages to underscore political persuasion and political and economic systems as determinants of environmental behavior.

Teddy assures her that there are plans in place to address the problem and that they must act immediately, to which Orlean responds, "Oh, hey. Alright, alright, alright. [Turns to her Chief of Staff:] When are the midterms?" She insists that this news must not break before the midterms because then "we lose Congress. And then there's nothing we can do about it anyway," to which Jason adds, "It'll be gridlocked." This conversation again highlights how politicians prioritize their political careers and thirst for power in their motivations over anything. But this is not all; this also plays into how a major party promises to take meaningful climate action, but when they are in power they do not act, citing reasons such as political fractions and partisan deadlocks. Ironically, they manipulate the same deadlock for stoking fear that if they are not reelected nothing can be done as the opposition does not care about climate change. This is eerily akin to the promises and threats of the Democratic Party in the US: they accuse the GOP of denying climate change and scare voters to vote for them but when in power they fail to deliver. Of course, they accuse the GOP for obstructing their agenda, and use the same strategy to persuade voters to vote for them in the upcoming election again.

The partisan diatribe is confirmed by the framed picture of Bill Clinton (Democrat) embracing Orlean. The Oval Office also features more images of notorious past presidents, including Andrew Jackson and Richard Nixon (Republican). We also see a framed picture of the President with Steven Seagal, an American-born actor and martial artist and a supporter of Vladimir Putin who was granted Russian citizenship in 2016. He served as Russia's special envoy to the US between 1996 and 2018, and has been the subject of several accusations of alleged sexual harassment or assault. And another picture of Orlean with Mariah Carey, an American singer, songwriter, record producer, and actress. Rather than a criticism of one or the other party, therefore, the movie decries the inefficacy of the two-party system. The mixture of Republican and Democratic politicians and celebrities is less to further stoke partisanship by favoring one party over another in the two-party system but is rather meant to underline Orlean's opportunism and questionable choice of role models as the President.

Despite Randall, Kate, and Teddy's insistence and the urgency of the situation, the President dismissively concludes that we should "sit tight and assess." Downplaying the significance of an extinction event, President Orleans adds, "Do you know how many, uh, the-world-is-ending meetings that we've had over the years? Economic collapse, loose nukes, car exhaust killing the atmosphere, rogue AI." Jason picks it up by saying, "drought, famine, plague, … alien invasion, population growth, hole in the ozone, …." Desensitization to apocalyptic disasters might explain why scientists' attempts to warn politicians fail, and instead of being alarmed and taking proportionate and timely action, they choose to sit tight. This is actually one of the potential negative impacts of over-using the fear appeal in communication: it presents the audience with a conundrum. Should they endure constant stress and anxiety or should they prioritize their psychological wellbeing and shut down on somber and apocalyptic messages? This communication strategy, therefore, risks pushing people to indifference and inaction. Besides, Orlean and Jason show that they perceive the threat posed by the climate crisis on a par with the problems they list. This underlines how the Administration has failed to truly grasp the extent and urgency of the climate crisis as an existential threat to life of all species on the planet.

Randall is outraged at Orlean's indecision and inaction, but she knows how to manipulate people. She manages to pacify Randall's outrage by a few words of praise that immediately pushes him to compliance. Teddy, however, discerns that the White House has decided to ignore them and the threat altogether and suggests that they should leak the information they have to the press. However, this is not enough to coerce the White House into action. Not only does the Administration refuse to change its behavior, it even denies having met them. However, Orlean suddenly decides to change course after a while as the scandal around her Supreme Court nominee Sheriff Wade Conlon exacerbates. It has also come to light that Orlean and Conlon have been having an affair for years. The media reports that the President has sent Conlon pictures of her private parts. Social media explodes with unqualified comments and memes, and members of her party set to drawing a formal censure of the President. She finds acting on the existential threat expedient and endorses it in an opportunistic U-turn for damage control.

To salvage herself from the scandal, President Orlean orders Randall, Kate, and Teddy to be arrested and brought to the White House where she greets them with "I think it was Winston Churchill or it was Bobby Knight [basketball coach] who said, 'The great labor of leadership is to lay down one's yoke of pride and admit that mistakes were made'." Besides her sheer ignorance, this also conveys her arrogance. The use of passive voice shows she is not even willing to admit she had made a consequential mistake. It is her son that hesitantly offers the scientists a Presidential apology only after deprecating their scholarship. While Randall is excited about this seeming change even after his unnecessary arrest, Kate proves to be more perceptive. She reacts by saying, "So, basically, what you're saying is you're about to lose the midterms 'cuz you got caught sending pictures of your cooch to your porn star sheriff boyfriend, so now it's to your advantage to act on the comet?"

Orlean's politically motivated opportunism does not know any limits. Her son turns to Dr. Calder and tells her she has to take the blame for their belated response. Dr. Calder concurs, saying, "Of course, Janie. I will resign in disgrace first thing tomorrow." Later, we see her in tears. Dr. Calder had downplayed Randall's and Teddy's pleas from the very beginning by saying, "Well, let's not be dramatic here. ... OK, let's all calm down." When asked to share the data with the scientific community, she had refused and insisted that she "serve[ed] at the discretion of the President." Later, we learn from a *New York Herald* journalist that she was instrumental in undermining Randall and Kate's efforts to reach out to the public by undermining the credibility of their claims, calling them "near-miss hysteria." Despite the services of this former anesthesiologist appointed the Director of NASA and a President Orlean super donor, Orlean does not hesitate even for a moment to sacrifice her career and reputation for her political future. In return for her loyalty, Orlean promises to have her back, and we later see her again working on the Bash project to reclaim the comet.

To recover from the Conlon scandal, the White House decides to drop a spectacular presidential statement on a warship. The message is aired in primetime, interrupting *Jackpot Fiancé*, a show similar to popular reality television dating game shows like *The Bachelorette*. Orlean manages to project herself as a great leader and uses the opportunity to leverage the political agenda of her party to win their favor back again. She insists that

they need a hero for this mission, although it is possible and recommendable for it to be accomplished automatically and remotely. Randall is baffled by this decision, but Teddy remarks, "Washington's always gotta have a hero." Orlean has asked a mercenary who shouts "get a little air into it, you lard-ass. You look like a goddamn bunch of pussies. You know what happened to that last pussy I worked with in combat? You ever seen inside a man's torso?" to crying children exercising on the White House lawn to pose as the hero in this mission. The presidential statement is concluded by "May Jesus Christ bless every single one of you, especially the honorable members of my own party. We will prevail," followed by an extravagant display of fireworks.

Besides expedient opportunism and gross indecency, politicians are berated for fostering, co-opting, and manipulating patriotism and heroism. Orlean's insistence that they need a hero for the mission that should be prominently featured on the spectacle they throw and Teddy's remark foreground how politicians promote and manipulate a sense of nationalism and patriotism in order to steer public opinion and win approval despite their actual intentions and actions. The fact that the hero is chosen to be a racist (see below) mercenary that abuses young children chastises the conflation of nationalism, patriotism, and militarization. In fact, the movie identifies and criticizes militarization as a major reason that inhibits meaningful climate action. Orleans' closing remarks in her statement also imply a subtle censure of religion and its instrumentalization in politics. Belief in solutions outside of human control is among the deterrences of taking climate action. People assume that religious forces will take care of what needs to be done, and subscribe to inaction. The Evangelical Yule (Timothée Chalamet) is a mouthpiece for their point of view. He expresses his skepticism about the catastrophe, citing "if God wanted to destroy the Earth, He would destroy the Earth" as his reason. The only justifiable place for religious sentiments must be reserved for when we do all in our material(ist) capacities. On the launch day of the NASA mission, we see an African American girl reciting from Psalm 23. We also repeatedly see people praying in churches, mosques, and temples. And when all missions to deflect the comet have failed and doom is certain, Yule prays again. Religion might help with consolation and religious beliefs must be respected, but we have caused the crisis, and we must take genuine and timely action to ameliorate it.

The glamor, spectacle, and hero cult of Orlean's shenanigan pave the way for her political triumph: they succeed in engaging people across social media platforms; President Orleans wins the midterm election in a landslide victory, and her approval rating soars to 99.1%; and even her scandalous nominee and lover is appointed to the supreme court. Randall is endorsed as a sexy scientist, and Kate continues to receive unfavorable attention on social media where conspiracy theories also proliferate. The opposition party joins the playground by accusing scientists of being Marxists and calling the comet a hoax, while the White House and Bash effectively manipulate political divisions to manage the crisis to their benefit.

The comet eventually gets close enough to the Earth to become visible by the naked eye. Many take to social media to encourage others to just look up and see it for themselves. Yet some still refuse to acknowledge the existence of the comet despite the conspicuous evidence. Meanwhile, in the too familiar incendiary rhetoric of US-American and populist politicians, Orleans incites her audience by saying, "Do you know why they want you to look up? … 'cause they want you to be afraid. … 'cause they are looking down their noses at you. They think they're better than you." Her incendiary remarks give birth to the "Don't look up" movement. In the fashion of populist politicians, she poses as if she belongs to the working and middle classes, and manipulates class struggle and inequality to mobilize her base against the elite. Her campaigns, gestures, baseball cap, and style is reminiscent of President Donal J. Trump and MAGA crowds, and her son's comments about her remind one of Trump's comments on his daughter Ivanka. Accusing the political elite, Trump also often manipulates political corruption to pose as a hero and a political outsider who will fight for the people to change the system (Cf. *South Park*, season 27, episode 1).

The movie shies away from explicitly exploring class struggle and potential solutions like degrowth or commoning. First, socialist attitudes and Marxian discourse are less welcome in the context of the US. Second, the movie also operates within the entertainment culture industry and relies on neoliberalism and capitalist economy for its market. It actually resorts to the same strategy Orlean employs: spectacle of the celebrity culture to advance its own vested interests. Nevertheless, there are subtle references to class struggle and materialist critique. Explicit classism resurfaces when Jason is speaking at an Orlean's rally. He says, "there's three types of American people. There're you, the working class; us, the cool rich; and then them. [Booing]. I'm sorry, but we need them. We need them, because you build us up to fight them. The working class, the lower." Like Orlean, he uses third person plural pronouns. The absurd and unnecessary loop Jason tries to explain sheds light on how players of political power pit one group against another and manipulate polarization as their raison d'être. They are satirized for their divisiveness and the way they create and sustain effigies of nemesis in order to justify their aggressive behavior, violation of human rights, and military expenditure, and environmental degradation.

Class struggle and division of labor resurfaces again in Jason's comments. While the Administration and its enablers continue stoking fear about immigrants as a distraction from the real issue, one single individual looks up and sees the comet at a rally. The crowd eventually understands they have been lied to. They throw bottles at the Chief of Staff and a senator who is a staunch supporter of Oreal, and in response Jason calls them "fucking *rednecks*."

*Don't Look Up*, therefore, gestures toward the intertwinedness of socioeconomic status and environmental behavior. Curbing consumerism is yet another concept that does not bode well with the movie's market as it relies on it for its profits. Nonetheless, it also pokes fun at consumerism. At the Bash launch, Jason remarks, "I've been noticing a lot of prayers recently for people during this time, and I commend that. Um, but I also wanna give a prayer for stuff. There's dope stuff, like material stuff, like sick apartments and watches, and cars, um, and clothes and shit that could all go away, and I don't wanna see that stuff go away. So I'm gonna say a prayer for that stuff. Amen."

Capitalism presumes infinite growth, creation of wealth, and accumulation of capital which require unbridled production and consumption. Production requires natural resources and labor; however, natural resources are finite. As they are depleted due to excessive extractivism, production, and consumption the linear trajectory of extraction, production, consumption, and growth cannot be sustained (hence circular economy). Curbing consumerism and degrowth have been proposed to address this paradox, but they seem to be inherently at odds with neoliberalism and market capitalism. Jason is a myopic buffoon-figure who fails to understand the gravity of the existential crisis that the comet/climate change poses. Meanwhile, he is savoring "dope … material stuff" and consumer products, instead of thinking of precious lives. Capitalism is depicted to be operating autonomously: it does not serve people and their wellbeing. Material wealth and its accumulation is not supposed to facilitate and optimize the reproduction, survival, and wellbeing of Homo sapiens and other species. The short-sightedness of consumer materialistic cultures, in other words, is held to ridicule.

Political resistance to the scientific message does not originate merely in partisanship, domestic political divisions, and class struggle, though. It also comes from entities that have no personal interest in the domestic politics of the US yet are not deniers. They simply feel betrayed and excluded from the benefits as Bash and the US Administration have excluded all other nations from the mission and the rights to the minerals of the comet. This nods to a major conflict of interest in limiting carbon emissions. Developed nations have long used inexpensive energy sources and emitted carbon into the atmosphere to build their infrastructures and consolidate their economic domination. Now that affordable fossil fuels are enabling growth and giving rise to new competitors, they demand that developing nations must stop using affordable energy sources (Ghosh, 2021; also see Fressoz, 2024). Vested interests and colonial legacies thus hinder trust building and global participation. The movie does not spare international organizations like the UN from its censure for their sheer inefficacy.

Meanwhile, the scientific trio relies on grassroot movements for resistance (Cf. Saito, 2024). Teddy calls for divesting Bash in demonstrations and occupations; Kate picks up graffiti art; and Randall leverages his media presence as a sexy scientist. But the last of their hopes vanish into thin air when a parallel deflection mission fails. Russia, China, and India were excluded from the rights for the minerals, so they launched their own deflection mission. Their mission could potentially save the planet; nevertheless, it is not motivated by the right reasons. Their actions are not primarily motivated by ethical considerations or concern for the planet's well-being, but rather by the pursuit of material gain within a competitive global (dis)order. After their mission fails, the Bash mission is the last remaining hope to save the Earth.

On the day of the Bash launch, Randall, Kate, and Yule buy groceries and go to Randall's home where Teddy joins them to have their last supper. June forgives Randall for his infidelity, and Yule proposes to Kate. Meanwhile, at the mission room, Peter and Orleans bail out of the room, leaving Jason and others behind as soon as it is clear that the Bash mission has failed. Next, we see them on a helicopter calling Randall to invite him on the escape ship that holds 2000 people. Randall turns down the offer and decides to spend his last moments beside people he loves. As the news of the failure breaks, we see cities in riot. Patriot New Network reports, "of course, there's only one story everyone's talking about tonight: topless urgent care centers." The scene cuts into what looks like a very big orgy. As disaster descends, Jack and Brie—the new anchors—decide to spend their last moment "drink[ing] and talk[ing] shit about people." And Commander Drask is exercising his second amendment right by shooting at the comet.

22,740 years into the future, the Bash spaceship lands on another planet and its passengers from Inter Bank, Texton Oil (mind the pun on Exxon), Vendel Lobbying, and GTW Equity, leave their cryo-chambers and set foot on the new planet. Only 58% of the chambers have been successful and many have lost their lives. It is on this new planet that Orlean meets her fate after being attacked and killed by a Bronteroc.

*Don't Look Up* underlines how political persuasion inhibits or encourages pro-environmental behavior. It further criticizes how politics is reduced to spectacle in US-American culture and politics. The political system is also shown to be among the determinants of pro-environmental behavior, particularly at the collective level. A malfunctional two-party system that prioritizes serving the interests of corporations, securing economic growth, and pivots on partisanship, allows a feckless President, whose selfishness frequently blunders the response of her Administration, to first ignore and then botch the response to an existential threat and compromise life on the planet.

## Media, Social Media, and the Celebrity Culture

*Don't Look Up* also criticizes the hurdles that the media, social media, and celebrity culture pose for genuinely engaging and addressing anthropogenic ecological mutations. When Kate, Randall, and Teddy are reporting their discovery to the President in the Oval Office, an episode of what Kate dubs the "freak show" unfolds. Orlean starts telling a "funny story" on finding her pack of cigarettes: "When I was running for President, I had to sneak cigarettes for the whole first month of the campaign, because, you know, photographs of me smoking went for, what, a hundred grand a pop?" But she was not able to hide it much longer and decided to smoke whenever she wanted. "Guess what happened? [triumphantly] I went up three points in the polls." To which Jason adds, "They loved that she kept it real." It is not honesty, decency, accurate representation, or verisimilitude that defines reality in a post-truth world; rather, it is mediated simulacra that creates reality. The media simulates a Baudrillardian Hyperreality that people endorse as reality (Baudrillard, 1994).

Orlean's enthusiasm for media attention and her manipulation of representations are strongly suggestive of Trump. These are of course coupled with other parallels including a picture of Dick Cheney, a trophy for 2015 Webby Award for the best short film series "Lady Biz," and shelves stacked with her own book, titled *How to Manage Your Money Even When You Have None*, reminiscent of Trump's *The Apprentice*, *Trump: The Art of the Deal*, bankrupt businesses, and his inflated claims to wealth. These parallels are further foregrounded by Orlean's son saying, "I can't think of another President that I've ever wanna see in *Playboy*." Besides his picture on the cover and interview in *Playboy*, this also resonates with Trump's comments about her daughter Ivanka.

After their first meeting with the President turns out to be a failure, the scientists are told by Jason that "obviously everything we talked about in here, super classified, right?" Nonetheless, they decide to leak the news to the press as they understand "the White House just blew … [them] off." Kate says that her boyfriend Phillip knows a reporter from the *New York Herald*, a thin disguise for the *New York Times*.

Phillip is Kate's boyfriend and a journalist representing the degeneration of ethical and moral principles in journalism. Discussing Conlon's scandal on phone with Kate, he says, "And get this. I found five former students who are willing to go on record that Conlon got noticeably aroused when he posed for their drawing class. Now, should I say, 'noticeably aroused' or 'engorged' because I don't want this to read like clickbait." Journalists do not hesitate to manipulate discourse and stories as clickbaits in the attention economy in order to secure financial gain in a rapidly changing media landscape. The irony is that he is concerned about a story that he can sell, a story that attracts the attention of the audience with sex and political scandal. Instead of sympathizing with his girlfriend who explicitly tells him she does not feel good, he shifts the conversation to his job and his petty concerns. When his unempathetic reaction fails to draw a response from Kate, he moves to passive aggressive behavior by changing the topic and claiming that Kate has an issue with his mom before we learn that his mom actually asked her if she "were a lesbian." Kate responds, "Hey. Can I sit down with your mom to have lunch in, like, seven months?" Phillip objects because "That's weirdly specific and distant." Phillip's mom suspects Kate to be a lesbian because of her looks including the dye of her hair and her piercings. Previously Jason has also called her, "Boy with the Dragon Tattoo." Later, he reaches out to fist bump her, sarcastically saying, "Thanks for dressing up." Her looks are also subject to public scrutiny when a staff crew is sent to help her with her "wardrobe," and Jack calls her "mullet girl" on the set of *The Daily Rip* show. Kate is an intellectual concerned over the existential threat of planetary extinction, while Phillip and others fail to comprehend the gravity of the situation, so they tend to ostracize and demonize her because of her different looks. This is one reason the date she offers to talk to his mom postdates the extinction event of comet collision.

Kate's boyfriend keeps complaining about not receiving her input about his article until she decides to tell him what the threat is about. On hearing the truth, he is petrified. He loses control, heads toward the street, and starts gasping. He is briefly convinced about the gravity of the situation, puts it above his professional development, and supports Kate in getting the story out. However, after Kate's first disastrous appearance on *The Daily Rip*, Phillip publishes an article in *Autopsy,* entitled "You Know the Crazy Chick Who Thinks We're All Going to Die? I Actually Slept With Her." Later, he tells a reporter that he is writing a book on his relationship with Kate, called *A Brush with the Devil.*

The staff and directors at the *New York Herald* are no better journalists than Phillip. The initial impression at *New York Herald* is professional journalism as they have made sure that they collect and verify pertinent information, but their meeting soon degenerates into a partisan attack and a media spectacle. They intend to protect the whistle-blowers by assigning them a pro bono attorney, but when Randall objects to this because he simply thinks he will be telling the truth to the public, one *New York Herald* representative scornfully says, "Make sure this one gets media training before he hits the shows. He seems a step slow." The magazine has arranged for them to appear on *The Daily Rip* morning show before their article gets published.

The role of the media moves to the next level as they arrive at the world of media simulations for the show. The crew sets out to give them a makeover, while Randall has yet another panic attack. Revamping their appearance is so powerful that even Randall's family are struck by how good-looking he is when they see him on their screen. As the show starts Brie Evantee (Cate Blanchett), one of the co-hosts, has a hangover and worries about the way her bosoms look. Randall's family, Kate's boyfriend, and Teddy are anxiously following the show. The show's headline is the unfolding story of the SCOTUS nominee Sheriff Wade Conlon, then the lineup will continue with Riley Bina's breakup story, and in the end the apocalyptic news. The media foregrounds and prioritizes trivialities while being unwilling to engage serious existential challenges.

When Randall and Kate eventually appear in the last segment, the first question they receive is, "Can I ask you something before we go any further? I have to know this. Is there life out there?" Randall manages to give an ingratiating response that persuades the hosts to like him, but Kate fails to strike the same chord by dumbing down or sugar-coating her message. She can no longer restrain her justified rage when the hosts continue trivializing the existential threat in their attempt to "keep the bad news light" and explodes into saying "maybe the destruction of the entire planet isn't supposed to be fun; maybe it is supposed to be terrifying and unsettling" before storming out of the studio. Her explosive yet founded reaction to the apocalypse triggers regret on the part of Teddy and *New York Herald* employees. Her boyfriend's companion who is watching the show with him tells him, "Hey man, my brother has bipolar disorder, so if you ever need anyone to talk …."

Kate's overtly emotional response on the show is analyzed on audience engagement charts at *New York Herald*. Audience engagement on social media peaks when Riley says she will take DJ Chello back; Kate's cursing on air also stirs some commotion among users, but engagement plummets with the backlash. Her justified horror and righteous anger is received and transformed into derisive memes on social media. Randall's complacency, coupled with his good looks, is received in a much more favorable light. He is now labeled AILF or "astronomer I'd like to fuck." The results of the audience engagement analysis indicate that "we did not feature or push on any sites, and clicks overall were below basic weather and traffic stories." Subsequently, the journalist on the story accuses Kate and Randall of the inaccuracy of their claims, although he had them fact checked in the first place. *New York Herald* subsequently decides to drop 'the story.' Tech companies and their algorithms foreground content that ensures user engagement regardless of its significance, validity, or morality, and the evolving profession of journalism has followed suit. Discovering the truth is no longer the objective in a post-truth world; engaging users on cyberspace and social media to increase the number of clicks and views they receive is (Cf. Srnicek, 2017; Varoufakis, 2023).

Besides social media, *Don't Look Up* also examines the negative impact of the culture industry and celebrity culture. The complacency of the entertainment industry is underscored with Hollywood's stance on the Just Look Up movement. *Total Devastation* is a movie produced on a budget of $300 million and is set to be released on the day that the comet will collide with the Earth. In other words, Hollywood fails to understand and engage the scale and gravity of the situation, instead treating it as yet another business opportunity. Even worse, it reproduces its distorted understanding with the great resources available to the industry and feeds it to its wide audience. The star of *Total Devastation* insists on taking a neutral stance in his interview, because he thinks that we need to stop arguing and just get along. The interviewer calls his gesture "so refreshing." The sheer fallacy and absurdity that climate change is up for debate or that you can possibly take a politically neutral stance toward it underlies the complacency of the industry and their priority in optimizing their profits.

Celebrities constitute a large part of what feeds the entertainment industry as well as media corporations. *Don't Look Up* explores the celebrity culture in the figures of Riley Bina, Brie, and Jack. Riley Bina (Ariana Grande) and the story of her breakup with her boyfriend haunts the movie. Surfing the internet for news and trends, Kate informs Randall at the White House that, "well, there's a bunch of stuff about their nominee not being qualified, but I don't see any articles about our thing." Teddy reacts by saying, "social media is just going on and on about that singer Riley Bina and DJ Chello [Kid Cudi] breaking up. That's just too bad. [Pause] They seem like sweet kids!" Even Teddy is drawn into the story. The audience is likely to interpret his expression of regret before the pause as resulting from the fact that the media and people are distracted by trivialities which stops them from engaging the real issue, i.e., the existential threat. When he finishes his sentence, however, one understands that he is also concerned about the breakup of these two celebrities.

The media also promulgates celebrity culture as it feeds on it. When the stage manager is briefing the hosts about the lineup for *The Daily Rip* show, Riley expresses her desire to talk about her new single, "a portion of [whose] profits goes to the manatee sanctuaries." The stage manager responds, "yes, but the breakup first." In the same room, Riley learns that Randall and Kate have discovered a comet, so she comments, "that's so dope! I have a tattoo of a shooting star on my back." Randall tries to be nice and reciprocate the kind comment with a short remark that he is sorry to hear about her breakup to which Riley responds, "why don't you mind your own business, you old fuck?" On the one hand, celebrities exhibit the details of their personal lives and toot their own horns as if everything revolves around them; on the other, they become defensive when people get curious about their personal lives. The media rewards the crude egocentrism, narcissism, and complacency of celebrities with comments such as "you're an inspiration. … You're really owning that pain." *The Daily Rip* show has staged a surprise call-in from Riley's ex-boyfriend and blatantly relies on melodramatic effects to attract the attention of its audience and capitalize on their personal life.

The audience is also complicit in the process. They melt when Riley says on air that she wants to take her boyfriend back "after many sleepless nights … three!" Melodramatic sentimentalism is immediately valorized by the ripple effect when social media explodes. Randall's son reports, "Bash News just sent me eight push notifications" and his brother says, "my phone just purchased DJ Chello's latest single … without asking." The sensational story climaxes with DJ Chello joining the show live and proposing to Riley. Kate's boyfriend, who is watching the show, responds to the charade with "come on! Ten famous second chances for two-timers …?" This underlines the less than healthy relationship between the celebrity culture, the media, and social media.

The impact of celebrities is of course not negative by default. Celebrities can also wield their social capital to mobilize their fan base. Later in the movie, Riley Bina organizes a concert to support the Just Look Up movement. Kate and Randall open the event with a call to the "nations of the world [to] defy Orlean and Bash. Launch your own nuclear deflection mission now and save this planet!" Nevertheless, the comments on the live stream of the concert are not all promising. The unequivocal message of the lyrics is to "get your head out of your ass/listen to the goddamn qualified scientists/… you can act as if everything is really all right/But this is probably happening in real time/… tomorrow may never come/Just look up/Turn off the shitbox news/'cause we're about to die soon, everybody!"

Brie Evantee and Jack Bremmer (Tyler Perry) are two more celebrities who are censured for undermining scientific evidence. Jack does not hesitate to tap into his appeal to populism when he asks Randall and Kate about aliens. His sheer ignorance is exposed in his mistaking Subaru car manufacturer to be the manufacturer of Subaru Telescope, named after the open star cluster known in English as the Pleiades. Brie pushes the limits even further in flirting with Randall on air. She approaches him immediately after his first appearance on the show and soon starts an affair with him. He relents to Brie's advances and starts an affair with her, despite the fact that they do not know anything about each other. Brie confesses that she "despise[s] the … getting-to-know-you part of relationships." She is a very shallow and egocentric character who emphasizes her credentials, sexual relationships (that include Brie sleeping with two former US Presidents), and her possession of "two Monets" when she opens up about her life to her lover Randall. Later, when she thanks Randall for "the most propulsive thing that has ever happened to" her, we see that the relationship for her was no more than a boost to her career.

Alongside the media and celebrity culture, social media is also identified as hindering constructive communication and engagement in addressing the existential threat. Social media is depicted to impede people from genuinely engaging with the problem and instead pacifying them into complacency. We see people hurt themselves in ridiculous challenges when the launch of NASA missions goes viral on social media with the #LaunchChallange.

Randall also quickly and steeply spirals down social media and trolling. He is somewhat thrilled at the prospects of featuring on many shows including *Sesame Street* and being adored as a sexy scientist. While social media fame distances Randall not only from his family but also from his cause, conspiracy theories proliferate online, eventually evolving into the political creed of Don't Look Up! The initial posts that the news of the comet gives rise to are only too familiar: "Jewish billionaires invented this comet threat so the government can confiscate our liberty and our guns."

Social media is not a distraction just for people and scientists; it also impacts politicians in a negative way. In the wake of the Bash mission failure, Orlean forgets to take her son with her to the escape ship. In the post credit scene, we see Jason appearing from under the rubble and debris as the only person who has survived the apocalypse. Coming from under the rubble, he is still holding on to the nuclear football. The first thing he does is to call for his mom. When there is no answer, he takes his phone and records a post: "What's up y'all? I'm the last man on Earth. Shit's all fucked up. Don't forget to like and subscribe. We out here." His obsession with social media runs so deep that he cannot help posting even as the last surviving man in the midst of total annihilation.

## Academic Institutions and Academics

Many have identified *Don't Look Up* as political satire, and I tried to show the working of satire in the movie. However, the movie tries not to reduce the problem to a partisan issue. Various stakeholders have failed to take timely and effective action on climate change partly because scientists and the media have failed in effectively communicating scientific findings and shedding light on the economic and political conflicts of interest that inhibit meaningful climate action.

For one, governments are not always eager to facilitate international scientific collaboration even when the global scope of the problem requires it. After Dr. Oglethorpe asks Dr. Calder to share the discovery of the comet with other academic institutions, for instance, she insists that the data is classified information, and she "serve[s] at the discretion of the President." Political, national, and parochial interests, therefore, can hinder scientific cooperation on issues that will inevitably impact every nation and all species inhabiting the planet, particularly due to the myopia of politicians and those in charge of mapping out national interests. The movie also comments on other political reasons behind the inefficacy of scientific findings including ignoring or muzzling scientists and compromising their integrity due to vested interests (for the complicity of scientists in the Necrocene see McBrien, 2016).

Besides political restraints, the failure of scientists to galvanize people, corporations, and decisionmakers has professional and personal reasons. *Don't Look Up* partly attributes the fact that environmental messages fall on deaf ears to academic institutions. This is obvious in Randall's and Kate's approach to communication. They both presume that data is enough to trigger changes in decision-making and behavior. However, the presentation of data is not enough to compel politicians, corporations, and people to take action. Both Kate and Randall focus on information provision and fear appeal while ignoring the power of imagination and visceral engagement for argumentation and persuasion (Boykoff, 2019; Zekavat & Scheel, 2023).

This myopia in academic thinking leads to academics putting all their hopes and assets in the much-cherished peer-review process. Randall, Kate, and Teddy repeatedly insist on the peer-review process as the golden standard of academic excellence and reliability. By contrast, the movie underlines how conflict of interest, among others, can undermine the peer review process. While gatekeepers tend to guard conventions and define trends and buzzwords, institutions continue to reward peer review publications rather than meaningful engagement with policymakers and genuine public participation. Peer review is not the only academic standard that the movie criticizes. It also pokes fun at academic institutions in general like repeatedly criticizing the prestige of universities as if they are corporate brands. The fact that distinguished scientists like Nobel laureates and Polonsky Prizes winners work for Bash and CERN Large Hadron Collider show how they sell out to power and money in enabling Bash to further its research agenda and undermining public good.

*Don't Look Up* further reflects on how academic institutions fail to prepare students and scholars for engaging and addressing real world problems. It criticizes the sheer inefficacy of academic training that fails to prepare students for serving the society and to effectively deal with and proffer solutions for anthropogenic ecological mutations in a way that can encourage people, policymakers, and corporations to act on those solutions. Kate, a PhD candidate, conveys the existential threat she feels by over-relying on fear appeal. She repeatedly underlines the dire consequences of the collision and tells her audience, "we're all gonna die." She fails to see the downsides of the apocalyptic narrative which effectively drives her audience to (eco)anxiety, causing indifference and inaction. The sobriety of her message regarding an existential crisis is therefore received as hysteria, and her image is metamorphosed into that of a demon in internet memes.

Knowing and doing better is much to expect from her as her supervisor does not show superior discretion or communication skills either. Randall fails to reach out to public, governmental, and corporate sectors and to policymakers, and is repeatedly said to need "media training." He fails to communicate the urgency of the situation to the President in the Oval Office. He tends to use jargon and rely on scientific terminology to underline the validity of his claim which cost him the understanding and attention of his untrained audience. When he tries to articulate the science, he is interrupted by the Chief of Staff who tells him "you're breathing weird. It's, uh, it's making me uncomfortable."

The fun that Jason pokes at Randall for the way he communicates his message is relevant to dominant environmental communication. In criticizing Kate and Randall's approach to communication, the movie does not only criticize the failure of academics to reach out to different stakeholders and communities, but also the general discourse of environmental communication. Climate change communication has largely relied on fear appeal and apocalyptic messages. In the same way that Kate and Randall fail to connect with their audience, this type of messaging risks losing the attention of people who are willing and try to listen but since the message remains too stressful, they cannot maintain their attention.

The movie provides a concrete example of how fear appeal can derail environmental efforts. After the NASA mission is aborted and Bash takes over the mission to extract rare metals from the comet, Randall, Teddy, Kate sink into despair. While they are discussing the situation in a restaurant, people overhear them and demand that they should share any information they have. Kate loses it again, and shouts how the government and Bash are putting everyone at risk for their own vested interests. Kate tells them that the government has decided to let the comet collide with the Earth "to make a bunch of rich people even more disgustingly rich!" Her message lands of course, but to a very different effect from what they desire. It does not start a constructive public debate, a mass movement, or any socio-political pressure that can lead to a change of course in managing the crisis. Rather, it causes street riots, violence, looting, and chaos. Inadequate communication not only distances us from achieving the cause but can also cause new problems and further complicate the situation. Dr. Oglethorpe seems to be the only scientist who is aware of the power of imagination and narrative as he recommends Randall should tell a story and "keep it simple. No math!"

## Personality and Environmental Behavior

The movie therefore criticizes the sheer inefficacy of academia, and by extension academic training, to effectively deal with and proffer solutions for climate change in a way that can encourage people, policymakers, and corporations to act on those solutions. Besides institutional failures, the movie is also cognizant of personality (e.g., mood, emotion, intelligence) as a determinant of environmental behavior.

Kate manifests several of these factors. Besides her problems with her journalist boyfriends, we see her reactions to be often informed by her mood and emotions. In the White House, Kate freaks out to the point of being sick at the thought of meeting the President to inform her of an event that threatens the planet and all species with extinction. Later, she confesses she is scared and cries five times a day. She grabs a Xanax pill from Randall and immediately swallows it when he tries to take one while waiting at the Oval Office to meet the President. She also resorts to substance abuse in response to the escalation of the events.

Personality as a determinant of environmental behavior is even more prominent in the case of Randall. Initially, he is not even ready to accept the situation and seeks refuge in denial, saying, "what the fuck? Is this real? Is this a God-damn joke? Holy shit. This can't be happening. No." When Kate freaks out at the thought of meeting the President, her supervisor fails to show empathy or give her comfort or assurance at her moment of vulnerability so she vomits because of stress. Randall does not seem to be very high on emotional intelligence. He clearly suffers from anxiety and panic disorders. We see him repeatedly taking Xana, hyperventilating in panic attacks, and repeating, "you're here now. You're here now. Eleven benevolent elephants. Eleven benevolent elephants." But it is not just his psychological infrailities that impede him; he is also depicted to be out of touch. When they board a military airplane in an emergency to go to Washington, DC, he asks, "is there any customary seating …?"

Randall's frailties lead to a failing marriage. When staying in a motel in Washington DC, he does not confide the unfolding events to June, his wife, on the phone. We see June being supportive and encouraging; she insists that Randall should speak up and take credit when meeting the President. Despite the love and support he receives from June and their two sons, Randall starts an affair with Brie. One night when Brie and Randall return, they find June, who suspected his betrayal, waiting for him in the room. Brie insists that they should skip the denial part and confesses that she has had an affair with Randall. For her, the question is whether Randall wants to stay with her or go back with June to Michigan. As Randall hesitates in his response, June breaks up with him, revealing his many problems including his panic attacks, depression, hypertension, restless leg syndrome, "and for America's sexists scientist, a bottle of goddam Cialis," a medication prescribed to treat erectile dysfunction. Brie of course is not interested in an emotional and meaningful relationship; she does not even care about Randall who fails to see this. Later on, Randall's disillusionment amounts to confessing to Brie at his arrest that he really thought he loved her to which she responds in puzzlement, "really!? Wow! That's wild." Brie of course does not hesitate even for a moment to ditch him.

We might initially get the impression that Randall upholds academic integrity. For instance, he is not willing to leak the news to the press after the White House ignores them. He says, "I just feel like this isn't what I do." It does not take long, however, before we see his academic integrity to be compromised. In the Situation Room on the launch day of NASA mission, Randall fails to acknowledge that it was actually Kate not him that discovered the comet and takes credit for the discovery. Later on, we see him identified as the discoverer of the comet in TV advertisements. After his rise to fame and his promotion to the position of the Chief Science Advisor to the White House, he loses focus and even ignores Teddy and Kate. Randall fails to preserve his integrity and make the right decision, and eventually succumbs to Bash and the Administration, even though he knows that the course of action they have taken will eventually lead to disaster.

He compromises all standards he cherished one by one. We see him in awe of big names and prestigious institutions, clearly stepping back from the 'golden' standard of peer review. He turns completely complacent when he starts to actively promote the Bash mission by appearing in their advertisements which are designed not to address doubts, provide answers or consolations, but to enhance Bash profits. Besides Randall as a scientist, the US Administration and FEMA are also complicit in this capitalist charade. He is eventually reduced to "a sort of nerdy sex object, who then gets co opted into government service to give the corrupt Administration the cover of scientific legitimacy" (Brody, 2022).

It is not just Randall that sinks into complacency. After an initial stormy start, Kate also descends when she is taken "off the grid" and threatened to stop interfering with the Bash mission. She simply surrenders despite her previously passionate and emotional reactions and returns home only to find that her parents are not willing to admit her into the house as they are "for the jobs that the comet will provide." Turned down by the media, politicians, the scientific community, her boyfriend, and parents, Kate subsides into indifference. She descends so low that she is even willing to turn a blind eye to shoplifting at the store in which she is working at the register.

Kate is salvaged after meeting Yule who invites her to hang out with him and his counter-culture friends. This encounter quickly develops into romance. Randall is ironically salvaged by Peter Isherwell. When Peter, Orlean, and Randall are meeting as a part of the Bash mission, Peter introduces Bash Explore and Acquire Drones as "mankind's savior" in another subtle reference to the role of (counter-)religion. He considers the machine as his child and himself as Chronos. Randall sheepishly questions if Peter is concerned about the success of the mission and mildly accuses him of muzzling scientists and ignoring scientific standards. He unintentionally offends Peter by calling him a businessman. Peter is cross and objects:

> "This is evolution of the human species. … You know that Bash has over 40 million data points on you, on every decision you have made since 1994, doctor? I … I know when you have colon polyps months before your doctor does. You got four or five at the moment actually. You know, they're not of concern, but I'd have a checkup as soon as you can. But more importantly than that, much more importantly than that, I know what you are, I know who you are. My algorithms have determined eight fundamental consumer profile types. You are a lifestyle idealist. You think you're motivated by beliefs, high ethical beliefs, but you just run toward pleasure and away from pain. Like … like a field mouse. … Our algorithms can even predict how you'll die. … I looked you up after we met. Your death was so unremarkable and boring. I can't remember the details, apart from one thing. You're gonna die alone. Alone."

Beside privacy and unrestrained and unregulated power of tech companies as discussed under the section on corporations, there is so much to unpack in this revealing passage. For Peter, people are consumers rather than individuals, and he can pigeonhole them into eight pre-existing categories. Individuality and personality are of no consequence to him. He identifies Randall as a hedonist. But most importantly the epiphany of dying alone shakes Randall so hard that he changes course. He eventually proves the algorithms to be wrong as Randall dies surrounded by his loved ones.

We see Randall appearing on *The Daily Rip* again after this incident, but this time he has an entirely different attitude. On the show, Jack tells him, "if Bash's stock is any indicator, then we don't have to worry about the peer review. It is going gangbusters," as an explicit connection between the world of media and capitalism. Randall, however, refuses complicity this time. Rather, he is outraged, shouting, "would you please just stop being so fucking pleasant? I'm sorry but not everything needs to sound so goddamn clever or charming or likable all the time. Sometimes we need to just be able to say things to one another. We need to hear things." This echoes what Kate had said in her first and last appearance on the show. Randall continues, "if we cannot all agree at the bare minimum that a giant comet the size of Mount Everest hurling its way toward planet Earth is not a fucking good thing, then what the hell happened to us? … We should've deflected this comet when we had the fucking chance, but we didn't do it. … And I'm sure many of the people out there aren't even gonna listen to what I just said 'cuz they have their own political ideology." Brie interrupts him to salvage the situation that she feels is getting out of hand on live TV, but this time Randall remains steadfast, shouting, "the President of the United States is fucking lying. … this whole Administration has completely lost their fucking mind, and I think we're all gonna die!" Later, he appears on *Sesame Street* again with the message that, "every single man, woman, and child on the planet is going to die. … Kids, listen to me. You tell your parents that President Orleans and Isherwell are sociopaths and fascists." Thus, his narrative changes to one of undefeated despair, rather than hope.

## Corporations

Capitalism and corporations in general and media corporations and tech companies in particular are among the targets of satire in *Don't Look Up*. Bash is the most prominent corporation featured in the movie. When the team of astronomers celebrate their discovery, they are listening to music on a Bash speaker, Marshall (Randall's son) has a Bash phone, and Randall himself uses a Bash TV in his motel room. Bash emblematizes technological progress, corporate greed, ethical infringements, dehumanization, and an unethical interference in governmental decision-making.

We are first introduced to Bash by an advertisement that parodies inspirational product launches and promotions of large technology companies. The voiceover admits that "sometimes, it can feel even too much;" nonetheless, tech companies profit off the over-supply and unnecessarily wide choices they offer their customers. Their product range creates a problem and then they launch and market new products to solve the problems they have caused in the first place. In the same fashion, Bash launches Bash LiiF that promises "life, without the stress of living," an inherently absurd claim. The advertisement features Sir Peter Isherwell (Mark Rylance), the founder and CEO of the company, and the "guy that bought the Gutenberg Bible and lost it." His character is seemingly a cross between Mark Zuckerberg, Steve Jobs and Elon Musk. Before his appearance on the stage, the announcer asks the audience to "remember to avoid direct eye contact, sudden movement, coughing, or negative facial expressions." This precaution sounds extreme and might raise serious doubts about his emotional stability. One can argue that this is to hint at his extremely low emotional quotient. His frozen facial expression, stiff posture, and outfit only add to this impression. His eccentricity again resurfaces in an emergency cabinet meeting at the White House. While a scientist is explaining how Bash plans to reclaim the valuable metals and resources in the comet, Peter keeps sniffing at her hair.

What is curious, though, is that despite his corruption and eccentricities, he has successfully created a fan cult around himself as the hall explodes into applause when he appears on the stage, and we hear many fans shouting, "we love you, Peter." He explicates his emotional deficiency in saying, "all of my life's work really, I see, has been driven by an inexpressible need for a friend who would understand and sooth me. And, and, you know now all those years of work have come to fruition with Bash LiiF." Criticism is targeted toward how corporations manipulate narratives in order to capitalize on everyone's problems, including (mental) health problems that require empathy and professional intervention rather than shopping for new consumer products. Bash, however, replaces the need for emotional connection and care with manufactured material need and subsequent satiation: "Our new Bash 14.3, uh, phone is fully integrated into your every feeling and desire without you needing to say one single word. … The Bash 14.3 phone, when set to the life setting, instantly senses my mood through blood pressure, uh, heart beats, and … ." He is interrupted by the phone which declares his vitals show he is sad and then automatically plays a video clip of pets, with a sing-over "you are my best friend, you make me happy" to which he reacts mechanically. It is ironic that the phone contradicts him in saying he is sad while a few seconds ago he had said that his work has finally ended his emotional deficiency. The irony simultaneously works on a second level: he is sad while he is promoting a Bash product that is supposed to put an end to people's emotional problems.

The target of satire, therefore, is the society, featuring the unquenchable greed of billionaires, that compels people to strive to overcompensate their emotional insecurities and deficiencies by compiling material possessions; a society in which billionaires do not hesitate to manipulate consumers and capitalize on their emotional problems. One way in which they can accomplish this is by breaching people's privacy. Bash phone shares user data and schedules therapy sessions in clear violation of both data privacy and medical ethics. Such intrusions make it possible for the data that both the director of NASA and the White House head of staff insist to be confidential to quickly fall into Isherwood's hands.

In another instance of such manipulation, we learn that the advertisement is designed to use scientific findings in order to manipulate consumers. While ignoring the young girls who appeared in the advertisement on his side, Isherwood asks his production team, "are you sure the video of the puppy on the rooster is optimizing our prepubescent sense memory consumer sector?" Bash is manipulating multimodal affordances and psychological insights in order to market cellular phones at young and susceptible targets. The advertisement, therefore, chastises humanoid billionaires, corporations, advertising and tech companies, and people who fall for these traps and subscribe to be good consumers in serving their techno-feudal lords.

Its technological and financial upper hand provides Bash with considerable political leverage to the point that it aborts the NASA mission to save the planet. When President Orleans eventually decides to take action on the comet to salvage her political career in the wake of the political scandal around her nominee for the Supreme Court, all resources are mobilized in a concerted effort to abort the crisis. A reporter reports, "not since D-Day have we seen such a mass mobilization of unions and resources for such a noble task." Commander Drask, the Washington hero, arrives at the launch pad with a Extravehicular Mobility Unit featuring a grotesque band that connects the pelvic area to the chest. When everything is ready for launch, Peter bursts into the Situation Room. We learn that "he is a Platinum Eagle level donor to the campaign, so he has full clearance." (Compare the notion of state capture in *Honest Government Ads* in Chap. 5, and Elon Musk under the second Trump Administration). As soon as he sees that the NASA mission is successful, he commands to talk to the President in private. Orleans returns with critical information about the mission, while people and the commander of the mission in NASA witness that the rockets are changing course in disbelief.

Everyone is kept in the dark until Peter later clarifies what has happened in a cabinet meeting in which Randall partakes as the Chief Science Advisor to the White House. The astrogeologists at Bash Cellular have discovered that the comet is rich with valuable minerals used in the production of electronic devices. These natural resources are being depleted due to overproduction; besides, "China has her big panda paw firmly on almost all the mines that produce these valuable rare Earth assets." The NASA mission, we learn, was aborted to secure Bash with exclusive access to trillions worth of assets that the comet will bring. The movie underlines how corporate greed and US-China competition in a unipolar world are exacerbating environmental degradation. Having exhausted all frontiers and conscious of the limits of growth, technocrats and capitalists resort to opening new frontiers in colonizing space in the Necrocene. "Space colonization would be the capitalist's and the environmentalist's dream: habitats would be self-sustaining and cost-neutral, using solar energy, cultivating their own crops, mining minerals on the moon or asteroids" (McBrien, 2016, p. 17).

Isherwell, of course, puts a positive spin on his greed and claims that "when these treasures from heaven are claimed, poverty as we know it, social injustice, loss of biodiversity, all these multitudes of problems are just gonna become the relics of the past, and humanity is going to stride through the pillars of Boaz and Jachin naked into the glory of a golden age … interplanetary, interstellar, intergalactic existence for the human race." Isherwell's absurd and unfounded claims are met with applause from the cabinet members. Even Randall falls for these fantasies, but Teddy recognizes this as "the pretty little bow they're putting around this line of bullshit," and Kate adds, "I bet they're gonna say freedom and puppies, too." Corporations make big and shiny promises, but they end up serving their vested financial interests and eventually aggravating the situation. This is inherent to capitalism that thrives on scarcity and disaster. Isherwell's promises and efforts resonate with contemporary attempts to colonize Mars by similarly unhinged billionaires. This is of course before the final post-apocalyptic revelation when we find out that the 'human race' constitutes only 2000 selected people who are granted seats on spaceships to escape the end of life on Earth. This brings up the question of environmental justice.

## Environmental Justice

*Don't Look Up* alludes to themes of environmental justice; however, it falls short in adequately portraying its intrinsic interconnectedness with ecological transformations and their broader implications. The Bash LiiF advertisement features several young girls, some of whom are not caucasian. This is a subtle criticism of how corporations capitalize on tokenism and pretensions to diversity to appeal to different market segments. This attitude is clear in Isherwell's reaction to one of the girls. When she asks him if she may say something, he bluntly responds, "no!" Later, when she tells him how she loves him backstage, he avoids even looking at her, let alone reciprocating the sentiment.

The most conspicuous gestures toward environmental justice surround the White House and its satellite institutions. We see at least two paintings of native Americans when Kate and Randall visit the White House. Later,

President Orleans jumps at the opportunity to use the image of Teddy as an African-American scientist to her political benefit. Such gestures do not genuinely promote diversity, equity, and justice. She also frames Kate as representing "the dissatisfied youth, and the mentally ill, and other marginal groups," crudely clumping together different historically marginalized communities. These are just instances of tokenism used by political and corporate opportunists to advance their own agenda. Tokenism, however, does not offer a genuine solution to structural injustices as in the case of racism in the US. When FBI agents shout at Teddy to arrest him, the African-American scientist responds, "okay. Now be cool. I just have more skin pigmentation than you, 'cuz your ancestors migrated to northern Europe and developed lighter skin so as not to get …."

Besides policing and surveillance, we see racism run deep in other power structures and institutions. Commander Drask, the Washington hero, addresses citizens of the planet in the NASA mission launch: "a big hello … to all those proud white folks working hard. … I also want to say hello to all the Indians out there. Both kinds. You know, the ones with the elephant and the ones with the bow and arrows. Hey, why haven't you guys ever teamed up? How cool would that be, huh? I also wanna say hello to all the gays out there." General Themes justifies his gross racist and homophobic remarks by saying he belongs to a different generation.

The movie, however, fails to further explore the implications of racism for environmental justice. The only hint is the criticism of how wealthy nations relocate environmentally hazardous activities to poorer countries. This is implied in the rumor that Bash got Orlean to pay the Chilean government US$90 billion to let the tsunami wash away the coast of the country as a result of the impact event.

## Conclusion

*Don't Look Up* employs satire to criticize politicians and political institutions, the culture industry, media and celebrity culture, scientists and academic institutions, as well as corporations for the impasse on taking timely and proportionate action on climate change. While satire reigns large, comedy and humor, that feature a lesser degree of indignation and judgment, are largely peripheral. Jason serves as a harlequin figure, and the lyrics of Grande's song are humorous, but the message is by and large delivered via satire. Comedy and humor seem to be assigned the duty of attracting attention, while satire delivers the core message. Since comedy and humor are usually perceived not to be serious, they are prone to undermining the gravity of the message. Besides, comedians are not scientists, so they are not perceived to be qualified to opine on scientific topics. Though categorized as a comedy, *Don't Look Up* does not limit itself to inciting mirth in its audience and avails itself to emotions like indignation and despair. In this way, the movie strives to maintain and underline the gravity and seriousness of climate change. The primary message, furthermore, is not that climate change is real. Rather than merely debunking denialism, the movie assumes the validity of scientific findings as its point of departure. Satire instead targets those who are responsible for taking serious and timely action on climate change yet fail to live up to their duties due to conflict of interest. In other words, the movie does not make any scientific claims, and instead advocates integrity in research, listening to responsible scientists, and acting on their recommendations.

Satire also acknowledges the reasons behind skepticism and mistrust in people. It exposes how politicians and corporations manipulate or co-opt environmental attempts for their own benefits while people only face the negative impacts of climate change and are expected to bear the brunt of green transitions. Bringing this into public attention can encourage debate and put pressure on corporations and policymakers, which can potentially cause change in a democratic political system. Satire lambasts corruption and shames political and economic power to underlie mistrust and acknowledge the concerns of citizens who are disheartened to act pro-environmentally.

Besides enhancing credibility, satire is also used to address ignorance, uncertainty, and mistrust through providing information. It exposes and censures the sources that spread or manipulate uncertainty by exposing the reasons and intentions behind stoking skepticism. Taking uncertainty about climate change as the incongruity on which it functions, satire effectively diminishes its significance (Apter, 2005, 2013; Chattoo & Feldman, 2020; Wyer & Collins, 1992). By providing information and addressing trust, satire also addresses denial and judgmental discounting, while engaging deniers head-on through shaming and lambasting.

In the realm of popular cultural narratives addressing climate inaction, *Don't Look Up* offers a critique of late-capitalist values by satirizing interconnected social, economic, and political systems such as wishful technofixes, extractive capitalism, and celebrity culture, which hinder effective action. The use of satire as a communicative tool allows the film to engage audiences as both observers and participants in the narrative. By placing scientists at the emotional and authentic core of the narrative, the film highlights the impact of misinformation and political ideologies on climate change responses.

Despite its calculated attempts, there are several drawbacks to the movie. Its political satire is deemed not to be direct and radical enough. Brody (2022) accuses *Don't Look Up* of a cynical and apolitical attitude. "The movie's built-in ambition appears to be simply crude demagogy, reaching an apotheosis in viewers who've just watched at home opening their poorly insulated windows and yelling into the street, 'climate change is real!,' without any demand of policy or awareness of actions that the need to slow it might demand." The movie concentrates on the role of corporations and policymakers in hindering climate action. Financial interest and power are correctly identified as two reasons behind the intention-action gap. This comes at the expense of reinforcing cognitive biases like affirmative thoughts and confirmation bias. In other words, people who think that the environmental impact of their lifestyle is nothing compared to that of corporations might get the message that their bias is correct and that even if they change their lifestyle, things will not really change. In other words, the movie might inadvertently encourage them to think that larger and more powerful institutions are responsible for climate change and for mitigating measures, while discouraging them from considering their immediate personal impacts.

The movie offers a critique of existing structures but stops short of imagining and proposing alternative realities or exploring the origins of anthropogenic ecological mutations. While it criticizes consumer capitalism and narratives of hope, for instance, it avoids exploring concepts such as multispecies commoning or undefeated despair. Although the film critiques academic institutions, it does not adequately address the complexities of (gendered and racialized) scientific knowledge production and regimes of truth. By focusing on the fictional stories of scientists, politicians, and media figures, the film neglects marginalized voices, alternative epistemologies, and indigenous knowledges, thereby continuing to prioritize perspectives from the Global North, even when these are satirized. It further fails to fully convey that "the university is not an ivory tower, sheltered and separate from the world of exploitation and dissent. Academia is as implicated as any other workplace in conflicts between labor and capital, resistance and hegemony" (Bogad, 2016, p. 253). In critiquing the status quo, therefore, the film nevertheless fails to envision an alternative (Doyle, 2022).

Another significant flaw in the narrative is its portrayal of a crisis that has been unfolding and disproportionately affecting the lives of marginalized communities and those in the Global South as a looming disaster yet to occur. As Chaudhary (2024) maintains, the film ideologically parallels a universal phenomenon with the differentiated reality of ecological mutations. It equates an impending crisis with one that is already happening, rather than acknowledging its immediate and ongoing impacts. In its attempt to critique apocalyptic narratives, the film inadvertently becomes one itself, shifting from a depiction of living in the end times to representing the end of times. This framing obscures the realities faced by vulnerable populations and fails to engage with the nuances of the unfolding complex ecological mutations.

*Don't Look Up* ultimately exemplifies the very system it seeks to critique, leveraging the mechanisms it condemns to market itself effectively. While the film criticizes the culture industry, it simultaneously secures the interests of Netflix, illustrating a paradox in its commercial alignment. The film's critique of celebrity culture is undermined by its own star-studded cast, which clearly serves market-driven objectives. This high-budget production fails to sufficiently nuance its role within the culture industry, while relying on pathos and anger as certainty emotions instead of ambivalence. The casting of Leonardo DiCaprio, previously celebrated as the 'sexiest man alive,' underscores its reliance on the same celebrity culture it critiques.

This underscores the workings of the culture industry, as critiqued by Max Horkheimer and Theodor Adorno in their seminal work, *Dialectic of Enlightenment* (Horkheimer & Adorno, 2016). The culture industry, as they argue, refers to the mass production of cultural goods under capitalism, where entertainment is transformed into a mechanism of social control rather than a medium for genuine artistic expression. Horkheimer and Adorno contend that standardized, commodified culture—manifested through film, music, and media—manipulates mass consciousness by promoting passive consumption and reinforcing the status quo. Instead of encouraging critical thought or individuality, the culture industry pacifies the public, converting them into compliant consumers who uncritically accept dominant ideologies. *Don't Look Up* serves as a contemporary illustration of these dynamics, highlighting the complexities and contradictions inherent in critiquing a system while simultaneously operating in it.

In a similar vein, *Don't Look Up* avoids engaging in more uncomfortable and nuanced discussions, deliberately sidestepping real political confrontation to prevent alienating segments of its target audience. The film largely refrains from generating feelings of shame or uncertainty among viewers, instead opting for an abstract and vague political stance that steers clear of real-world politics. As Brody (2022) observes, "what Hollywood people don't want to do is to critique a political party, because there goes half the potential box office." Aware that political persuasion significantly influences behavior, *Don't Look Up* critiques the complicity of the entertainment industry and celebrity culture, yet relies on the same tactics to ensure its own financial success. It condemns consumer capitalism but remains unwilling to jeopardize its box-office returns. These hesitations and strategic considerations are evident in the topics the film excludes and fails to address adequately, such as gender, colonialism, racial justice, imperialism, and multispecies environmental justice. As Brody (2022) aptly notes, "its civic ambitions reflect, above all, the inside-Hollywood tunnel vision that it mocks."

Therefore, the normative and conservative ideologies that underpin Hollywood adventure narratives continue to dominate. These narratives tend to favor Western regimes of truth, which often resist challenging established binaries and hierarchical relationships. This Eurocentric ideology, typically centered on white, heterosexual, middle-class male protagonists who save their nuclear families from various threats, is deeply ingrained in the structure and ideology of adventure films. Falling back on these frameworks are particularly problematic in the context of climate change narratives, where advocating for a radically new direction for our planet's future is essential. As Salmose (2018) notes, Hollywood adventures, despite their global popularity and commercial success, largely fail to contribute meaningfully to the discourse on the Anthropocene. The genre frequently overlooks the necessary transition from the Anthropocene to the posthuman condition, rendering any potential sensational warnings it offers transient and superficial.

Thus, one might argue that the movie mobilizes ingrained biases and prejudices to monetize and capitalize on climate change, much like Bash or the carbon offset industry. By doing so, it risks perpetuating the very issues it seeks to critique. Potential remedies for this include adopting a self-reflexive approach to transcend the limitations of the culture industry (Zekavat, 2019, 2025). Besides, embracing uncomfortable and ambivalent emotions could enhance visceral and cognitive engagement on the part of the audience. This approach could help shift the focus from merely capitalizing on sensationalism to fostering a nuanced and more meaningful discourse. Lastly, prioritizing emancipation over justice could open up new possibilities for the Exhausted of the Earth to externalize ecological mutations and their intersecting and compounding issues, offering pathways for empowerment and resistance beyond traditional frameworks.

## References


Apter, M. J. (2005). *Personality dynamics: Key concepts in reversal theory.* Apter International.

Apter, M. J. (2013). Developing reversal theory: Some suggestions for future research. *Journal of Motivation, Emotion, and Personality, 1*(1), 1–8. https://doi.org/10.12689/jmep.2013.101

Baudrillard, J. (1994). *Simulacra and simulation.* University of Michigan Press.

Bogad, L. M. (2016). *Tactical performance: The theory and practice of serious play.* Routledge.

Boykoff, M. (2019). Creative (climate) communications. In *Productive pathways for science, policy, and society.* Cambridge University Press. https://doi.org/10.1017/9781108164047

Brody, R. (2022, January 6). The crude demagogy of don't look up: Adam McKay's star-studded allegory of climate change has a cynically apolitical view of politics. *The New Yorker.* https://www.newyorker.com/culture/the-front-row/the-crude-demagogy-of-dont-look-up

Chattoo, C. B., & Feldman, L. (2020). *A comedian and an activist walk into a bar: The serious role of comedy in social justice.* University of California Press.

Chaudhary, A. S. (2024). *The exhausted of the Earth: Politics in a burning world.* Penguin.

Doyle, J. (2022). Communicating climate change in 'Don't Look Up'. *Journal of Science Communication, 21*(05), C02. https://doi.org/10.22323/2.21050302

Fressoz, J.-B. (2024). *More, more, more: A history of the world's excesses.* Penguin Books.

Ghosh, A. (2021). *The nutmeg's curse: Parables for a planet in crisis.* University of Chicago Press.

Horkheimer, M., & Adorno, T. W. (2016). *Dialectic of enlightenment.* Verso.

Little, H. (2022). The use of satire to communicate science in Don't look up. *Journal of Science Communication, 21*(05), C06. https://doi.org/10.22323/2.21050306

McBrien, J. (2016). Accumulating extinction: Planetary catastrophism in the Necrocene. In J. W. Moore (Ed.), *Anthropocene or capitalocene? Nature, history, and the crisis of capitalism* (pp. 116–137). PM Press.

McKay, A. (2021). *Don't look up [film].* Netflix.

Saito, K. (2024). *Slow down: The degrowth manifesto.* Astra Publishing House.
Salmose, N. (2018). The apocalyptic sublime: Anthropocene representation and environmental agency in Hollywood action-adventure Cli-Fi films. *The Journal of Popular Culture, 51*(6), 1415–1433. https://doi.org/10.1111/jpcu.12742
Srnicek, N. (2017). *Platform capitalism.* Polity Press.
Varoufakis, Y. (2023). *Technofeudalism: What killed capitalism.* Penguin.
Wyer, R. S., & Collins, J. E. (1992). A theory of humor elicitation. *Psychological Review, 99*(4), 663–688. https://doi.org/10.1037/0033-295X.99.4.663
Zekavat, M. (2019). Reflexive humour and satire: A critical review. *European Journal of Humour Research, 7*(4), 125–136. https://doi.org/10.7592/EJHR2019.7.4.zekavat
Zekavat, M. (2025). The paradox of political satire: Navigating critique in culture industry and neoliberal media. *Comedy Studies.* https://doi.org/10.1080/2040610X.2025.2509429
Zekavat, M., & Scheel, T. (2023). *Satire, humor, and environmental crises* (1st ed.). Routledge.

CHAPTER 4

# Late-Night Shows and the Quandary of the Culture Industry

Television, as a major actor under the culture industry, exemplifies the commodification of culture, where entertainment becomes a vehicle for both mass consumption and subtle ideological dissemination. Late-night shows serve as prominent examples of this phenomenon, employing satire to engage audiences. These programs have a wide reach and multimodal appeal, utilizing not only textual and visual elements but also humor, celebrity presence, and music to captivate viewers.

Despite their occasional satirical nature, late-night shows are intricately tied to commercial imperatives. They generate revenue through advertisements and sponsorships, which can compromise their critical potential. Their working as an integral part of the culture industry was recently foregrounded when CBS announced the cancellation of *The Late Show with Stephen Colbert* in July 2025, effective May 2026, citing financial losses despite Colbert's consistent ratings dominance. While CBS attributed the decision to declining ad revenue and broader cuts across its late-night programming, many observers have highlighted the political undertones of the move. Notably, the announcement came just days after Colbert publicly criticized Paramount's US$16 million legal settlement with Donald J. Trump, labeling it a "bribe." This timing, alongside Paramount's pending merger with Skydance Media—which requires US Federal Communications Commission (FCC) approval—fueled widespread



M. Zekavat, *Leveraging Satire for Environmental Advocacy*, https://doi.org/10.1007/978-3-032-04838-7_4

speculation that the cancellation was a calculated move to ingratiate CBS's parent company with Trump and his allies.

The abrupt termination of Stephen Colbert's show sparked widespread criticism, with observers arguing that it heralded not only increasing suppression of dissenting media voices but also the growing willingness of corporate networks to capitulate to political pressure in pursuit of regulatory favor. These concerns were further substantiated when ABC suspended late-night host Jimmy Kimmel indefinitely in September 2025 following his remarks about the killing of right-wing influencer Charlie Kirk that were widely perceived as innocuous. The suspension came after FCC Chair Brendan Carr—a Trump appointee—publicly threatened ABC, seemingly using Kirk's death as a pretext to intensify efforts to cancel Kimmel's program. While the show ultimately returned amid mounting public pressure—including protests outside the recording studio, Hollywood advocacy, and a surge of Disney subscription cancellations—these events underscore both the escalating threats to free speech and the ideological role of commodified late-night programming within the broader culture industry.

While infotainment—entertainment that also informs—offers a platform for audiences who might not otherwise engage with traditional news media, it often falls short of advocating radical societal changes. Instead, these shows tend to remain complicit, providing complicitous critique within the bounds of what is commercially viable and palatable to a broad audience (Hutcheon, 1988). In other words, late-night shows reflect the dual role of television and streaming under the culture industry: as a medium of both engagement and commodification. They highlight the tension between cultural productions as a form of social critique and their function as monetized products tailored to sell other commodities, ultimately reinforcing the status quo.

Recent developments have significantly altered the landscape of late-night television and its role within the culture industry. The advent of the internet and the subsequent decline of traditional TV viewership have reshaped revenue models, as advertisers increasingly prioritize social media platforms over network television. In response, many late-night shows have expanded their presence online, uploading content to platforms such as YouTube and making episodes available through various streaming services.

This digital migration has facilitated new forms of appropriation. Both the shows themselves and individual users can excerpt segments, repurpose them for different contexts, or manipulate them for diverse objectives, whether constructive or detrimental. The interactive nature of Web 2.0 has enabled the content of these shows to be (re)appropriated, potentially serving more transformative purposes. The capacity for users to remix and share content allows for the possibility of extending its impact beyond the original broadcast, fostering dialogue that may challenge existing narratives or highlight underrepresented perspectives. While traditional media still plays a significant role, therefore, the participatory and decentralized nature of the internet allows late-night content to be leveraged in ways that can either reinforce or disrupt prevailing ideologies.

Late-night shows also exhibit a nuanced relationship with their audiences, which varies based on network demographics. They address a wide political spectrum from the far right as in *Gutfeld!* (Fox) to the left of the center, possessing a unique capacity to engage a broad audience, including millions of potential voters with politically-centric opinions. This is particularly significant in the context of anthropogenic ecological mutations, which have been sources of significant conflict since the latter half of the twentieth century (see Hendrix et al., 2023).

Climate change, as an overtly politicized issue, has often polarized public discourse. Historically, right-wing conservatives have either denied the existence of climate change or expressed skepticism towards it. Recently, there has been a concerning shift in their rhetoric, as they increasingly weaponize climate change to selectively manipulate public fears and advance their political agendas. This includes stoking xenophobia and waging ideological battles against climate refugees (e.g., Milman, 2021; Walker, 2021).

Beyond political polarization, climate change represents a battleground for conflicting interests: the overwhelming scientific consensus versus denialism, the priorities of a privileged few over the needs of the majority, and the tension between indulgence and abstinence. These conflicts are exacerbated by ineffective communication across sectors, often relying heavily on fear appeal. However, fear alone is insufficient to engage all stakeholders and demographics. Humor and satire, as employed by late-night shows, offer promising alternative strategies for environmental communication. These strategies can transcend traditional barriers of climate communication, fostering engagement through more relatable and less confrontational means. By leveraging their broad reach and capacity for nuanced critique, late-night shows can contribute to a more informed and engaged public dialogue on climate change.

Thus, late-night shows represent an intriguing case within the culture industry. Ideologically serving the culture industry, they sometimes manipulate psychological insights to protect the interests of the wealthy and powerful by shifting the blame for environmental degradation onto individuals or downplaying the environmental urgency. Their environmental messages sometimes emphasize personal and individual responsibility rather than addressing systemic issues. On the other hand, these shows also possess the potential to broaden their outreach, transcending political divides and engaging audiences who might otherwise remain indifferent to environmental concerns. The segmentation of late-night content and its potential for appropriation can offer a multitude of more radical opportunities in the age of social media. By leveraging satire and humor, therefore, these shows can critique and shame policymakers and corporations, advocating for change at both individual and collective levels.

In this chapter, I aim to demonstrate how satire and humor can be employed to hold powerful entities accountable and inspire action. By analyzing selected episodes of *The Late Show with Stephen Colbert* (CBS, USA), *Last Week Tonight with John Oliver* (HBO, USA), and *ZDF Magazin Royale* (ZDF, Germany), I will explore the affordances of humor and satire for environmental advocacy across cultural and linguistic boundaries and reveal how they employ psychological mechanisms to frame their messages. These three late-night shows have been chosen not only for their diverse linguistic and cultural contexts but also for their extensive outreach. They air on major networks, are available on streaming services, and are accessible for free on platforms like YouTube, granting them a diverse and international audience. Moreover, these shows are critically acclaimed, having received numerous accolades and nominations.

The format of *The Late Show with Stephen Colbert* enables the host to introduce a topic, joke about it, and move to the next topic. As they are best understood in isolation, several segments with environmental themes from recent episodes are selected for discussion. The format of the other two shows, by contrast, allows for an in-depth analysis of a single topic. I will start with *The Late Show with Stephen Colbert* as it follows the traditional format of late-night shows and satirical news. Each episode features a monologue in which the host humorously and satirically comments on the day's news without delving into analysis. *Last Week Tonight with John Oliver*, however, adopts a different format bordering on investigative journalism in that each episode largely focuses on just one piece of news or a socio-political topic; as a result, the comments tend to be more focused and nuanced. *ZDF Magazin Royale* largely owes its format to *Last Week Tonight with John Oliver* and will therefore be discussed afterwards.

## *The Late Show with Stephen Colbert*

*The Late Show with Stephen Colbert* is a talk show that has aired on CBS in the US since 2015 at around 11:35 PM EST and streams on Paramount+. Colbert uses humor and satire to advocate environmentalism at both individual and collective levels while providing information, inciting uncertain emotions, diminishing the scientific uncertainty about climate change as abused by deniers, and addressing the social determinants of environmental behavior.

The show frequently employs humor to draw attention to and provide information about climate change. In one episode, for instance, Colbert insists that climate change is real and anthropogenic. By attributing drought, storm, record-breaking precipitation, and strong winds to human activities rather than "bomb cyclone," he creates an incongruity that not only originates humor but also diminishes the uncertainty of anthropogenic climate change (Apter, 2013; Wyer & Collins, 1992; Zekavat & Scheel, 2023, "Rainfall Swamps The West, And GOP Members Conspired With Jan. 6th Riot Organizers" https://youtu.be/TiBtWFOLLjw). In another episode, he reports that the planet is getting a black box to record scientific data as well as news stories in anticipation of a climate catastrophe and annihilation of civilization. Assuming the perspective of future inhabitants of the earth, he says, "it looks like past humans ignored global warming until it was too late, never invested in green energy, and most of them killed themselves doing something called the milk crate challenge" ("U.S. Warns Russia Not To Take Ukraine|Earth's Black Box Will Record Humanity's Demise" https://youtu.be/QjEEiPGs9YE). The incongruity of what concerns people and what ought to concern them in the reference to popular culture, more specifically a TikTok challenge that had gone viral at the time, strives to shed light on inconsistencies in our priorities. It also creates an effective bathos only to enhance the humorous effect. Similar to the pain some social media users irrationally inflict upon themselves in hurting themselves while climbing crates, our collective environmental (in)decisions and (in)actions will come back to haunt us in the future. Colbert, therefore, seems to be using humor as effective scientific communication to "engage and educate less interested and knowledgeable citizens on important topics" (Yeo et al., 2022, p. 593).

The show seems to be aware of and tries to mitigate one of the drawbacks of using humor for environmental messaging. The identity of the communicator has been shown to impact the audiences' perception of the effectiveness of the message (Yeo et al., 2022). Since the host of the comedy show is not known for his credentials or scientific output, he resorts to citing scientific studies as a way to boost the communicator's credibility. In other words, scientific findings serve as a point of departure for foregrounding incongruities in our attitudes and behavior, hence the origin of humor. In one segment, the host reviews a piece of scientific news concerning how scientists can now decipher the meanings of the sounds pigs make. They can interpret each grunt by associating it with possible emotions. Colbert frames this scientific discovery as dark humor to shock his audience into reconsidering their lifestyles. He ironically says, "so pigs are capable of real, complex emotions. Well, it's nice to know my bacon understands how happy it's making *me* [emphasis added]." In this way, he frames a potentially environmental message in terms of ethics and animal welfare. So instead of dwelling on the carbon footprint of meat consumption, he uses empathy in order to encourage his audience to reconsider their diets. To avoid sounding arrogant and looking down on his audience, the joke employs a singular first-person pronoun. Reflexive humor, where the humorist and the target of the joke are the same person (Zekavat, 2019a), is used to eschew depicting an overtly gory image or transpiring a sense of complacence on the part of the humorist. Nevertheless, the gasp of his studio audience shows that he succeeds in foregrounding the suffering of a sentient being that has to lay its life to provide meat for our diet. Therefore, our egoistic values are exposed and pro-environmental injunctive norms are encouraged through shame (see Bore & Reid, 2014; de Groot & Thøgersen, 2018; "Meanwhile... Elon Musk And Grimes Welcome Another Child|Metallica's Lars Ulrich Is Now A Toilet" https://youtu.be/q8zlh8XKfLc).

Information provision is one of the ways to approach denialism through addressing uncertainty, mistrust, and reactance, and humor is a communication strategy for providing information, so it can be used to address denialism. Humor can also divulge the faulty logic of deniers through ridicule. Reacting to deniers of climate change, Colbert poignantly says in one episode, "it doesn't matter what you believe; it's true. That's like not believing humans cause pants pooping. It's just a natural part of the cycle of ma khakis. I'm gonna leave them for my grandchildren to deal with." Here he uses an incongruent, scatological analogy not only to create humor but also to reveal the incongruity of the excuses that the deniers of climate change might present. Ridiculing the deniers who treat anthropogenic climate change as if it is a matter of belief rather than a scientific fact simultaneously implies a sense of superiority. Besides information provision and exposing the flaws of argument in the discourse of denialism, this joke also highlights the disproportionately negative impacts of climate change on different generations. Since the catastrophic impacts of climate change are rather distant, older generations might be less willing to take action to address them as it is less likely that they will be directly impacted by these consequences. In other words, this joke underlines age as a determinant of environmental behavior (Zekavat & Scheel, 2023; "Food And Gas Prices Spike, And Rep. Mo Brooks Wore Body Armor At 'Stop The Steal' Rally" https://youtu.be/RPrTiLtqXVM).[1]

The last two jokes partly aim to elicit shame, an ambivalent emotion that can boost the impact of the message. Recent scholarship has conveyed that the sense of certainty that emotions elicit are more relevant to their impact on the depth of processing as compared to the positive or negative mood they create.[2] In other words, it is less relevant for a message to create a positive mood than for it to invoke an uncertainty emotion for impacting behavioral change. The use of humor and particularly satire stimulate strong ambivalent emotions that can mitigate the adverse impact of positive emotions on persuasion. Using humor and satire that invoke shame as an emotion that elicits ambivalence can therefore be an effective strategy to heighten elaboration and increase the impact of the message

[1] For more instances of the use of humor to provide information see "Climate: Changed—Is The Sun God Angry At Us?" https://youtu.be/2I_K63hOgzk; "Stephen Colbert's Dil-Do's & Dil-Don'ts" https://youtu.be/Pzgtt_QkIcE; "Uh-Oh, Water Is Now A Commodity—Stephen Colbert's Most Unfortunate Segment" https://youtu.be/gnEDF0tTReI)

[2] Emotions feature different cognitive dimensions like "the level of responsibility, pleasantness, control, and confidence they elicit" (Strick, 2021, p. 11) and can therefore be categorized according to the level of confidence they elicit. For instance, "emotions like anger, disgust, contentment, and happiness are all accompanied by a sense of certainty, even though the first two are negative and the latter two are positive. Emotions like worry, shame, surprise, and hope are usually accompanied by a sense of uncertainty" (Strick, 2021, p. 11). Humour and satire stimulate uncertain emotions and propagate ambivalence which in its turn can lead to strong attitudes predictive of future behaviour. For more on this see Zekavat and Scheel (2023).

(Skurka et al., 2018; Skurka et al., 2019; Strick, 2021; see also Norgaard, 2011; Panda, 2020; Seymour, 2018).

So far, I have tried to show how Colbert uses humor and satire to draw attention, provide information, foster empathy, chastise denialism, and attribute responsibility at an individual level as psychological mechanisms. The scope and complexity of climate change, however, require both individual and collective actions. Besides citizens, policymakers, organizations and corporations are major stakeholders in mitigating ecological mutations. *The Late Show with Stephen Colbert* is careful not to constraint its scope to individual environmental behavior. In a segment, for instance, Colbert starts with addressing American citizens but continues to say, this week "hundreds of world leaders are meeting with climate activists to attempt the one thing to fight climate change that no industrialized nation has done before: anything!" So he quickly shifts the focus from citizens to policymakers and continues to criticize them for their inaction.

He continues to return to age as a determinant of environmental behavior in saying "… ordinary people are doing something about climate change: they're worrying; especially young people. A recent study asked youth 16–25 from around the world how they felt about climate change, and 56% agreed with the viewpoint that humanity is doomed. Nice try kids, but you're not getting out of your student loans." Citing published studies to enhance the credibility of his argument, he underlines that younger people seem to care more about climate change and the future of the planet, while chastising the American public for their indifference or denial. His humor uses age, particularly aging politicians, to bridge individual and collective aspects of behavior. He appeals to the elder citizens by mocking how they obdurately choose to ignore the scientific evidence: "Americans treat climate science like soccer. We know it's out there and really matters to the rest of the world, but no one can make us care." From generational disparities at the individual level, he segues to the collective level when he says, "look no further than the US congress where 25% of the House of Representatives and 30% of the Senate refuse to acknowledge human-caused climate change. I'm surprised so few senators believe it. I mean Chuck Grassley[3] must be shocked by all the changes to the earth he's seen since his childhood in Pangaea.[4]" He forges ahead to say, "in

[3] An old Iowa senator born in 1933.

[4] A supercontinent that incorporated almost all the landmasses on Earth some 299 million to about 273 million years ago.

fact, our senior leaders' senior moments could be behind our slow reaction to climate change. The average age of US senators is 64.3 years, the oldest in US history. And a recent study found among Americans over 65, only about half are very or somewhat concerned about climate change. They're much more concerned about getting to dinner. It's already 4:15. What are we waiting for?" Having fully laid out the generational conflict that impedes taking meaningful and timely action to address climate change, he appeals to older citizens by saying that doing something about climate change might as well be a way for them to connect to their grandchildren.

In an attempt to make the consequences of climate change more relatable to the US-American audience, he cites a study that warns about the impact of climate change on crop failures, and singles out durum wheat, the main ingredient of most pasta products. In exploring the impact of climate change on a popular dish and common staple food, the show again manages to breach the physical and psychological distance of the risks of climate change and brings the immediacy of the threat closer to home. In other words, by bringing the problem home the show employs a potential strategy to counter judgmental discounting that causes the temporarily or spatially distant risks to be underestimated. Climate change is usually perceived to be a distant risk, therefore, framing it in familiar and immediate terms allows environmental humor to increase the salience[5] of climate change among its audience (see Chattoo & Feldman, 2020).

Maintaining the momentum of the immediacy and salience of climate change in the US, Colbert continues to discuss meat consumption in the country. He comments that meat "tastes so good, you'll never picture the animal's face." Referring to animal suffering, Colbert seeks to incite shame, call for a revision of egoistic values, and reinforce pro-environmental injunctive norms. The host, though, is fully aware that a short late-night show segment or a few jokes cannot reform the long cherished and deeply ingrained carnism. Besides, climate change, especially as it relates to the culinary culture in the US, has proved to stoke political conflicts. At the introduction of the Green New Deal, many Republicans, for instance, attempted to galvanize their base by claiming

[5] In psychology, salient means "distinctive or prominent. A salient stimulus in a multielement array will tend to be easily detected and identified. The noun form, salience (or saliency), denotes a parameter of a stimulus that indexes its effectiveness" (https://dictionary.apa.org/salient)

that, "... they [i.e. Democrats] want to take away your hamburgers! This is what Stalin dreamt about but never achieved!" (Remnick, 2019). So Colbert adroitly leaves the topic after mentioning it, and instead of dwelling on meat consumption amongst US-Americans, therefore, he shifts the focus to the negative environmental impact of the meat that pets consume in the country. In other words, the show selects a more feasible option to further its environmental agenda in advocating more sustainable protein sources for pet food and promoting insect-only pet food companies.

One reason behind this compromise is the intention-action gap. More aggressive measures are definitely necessary to tackle climate change, while overwhelming individuals can lead to despair and eco-anxiety, inadvertently driving them to inaction. Colbert explicitly refers to the intention-action gap when he says, "OK, I want to fight global warming, but that's going too far. I will not, not, give up my air conditioning." There are two explanations for the breach between our intentions to take action and our failure to do so. First, tokenism, that is to say, people prefer to take low-cost environmental actions with little mitigating effect rather than high-cost actions that might really make a difference. Second, individuals responding to a threat are likely to rely on one action, even when it provides only incremental protection or risk reduction and may not be the most effective option. After taking one action, people tend to stop taking more actions, presumably because the first one succeeds in reducing their sense of worry or vulnerability. This is usually referred to as single action bias (Shome & Marx, 2009; "Aging Politicians And Meat Eaters Stand In The Way Of America's Climate Progress" https://youtu.be/YcFlDnxiBiM).

While the previous episode tries to simultaneously call for both individual and collective action, some episodes put a greater emphasis on the political determinants of environmental behavior. Commenting on the news that China and the US, the world's two biggest greenhouse gas emitters, have agreed to "enhance ambition" to fight climate change, for instance, Colbert fustigates the phrasing of this diplomatic statement and the inaction of large polluters as emblematized in their absurd "diplomatic level of crisis response." Addressing the political aspects of climate change mitigation as in the case of US-China rivalry does not mean that individual responsibility is ignored. While foregrounding the political dimension of tackling climate change, Colbert manages to simultaneously underline the role of individual citizens in the process. The studio audience responds with a gasp when Colbert cracks a dark joke that reflects on both political and individual aspects of environmental behavior. After ironically aligning "any Keurig [coffee] machine" with the US and China as the three worst polluters, Colbert mordantly says, "I'll take my coffee with milk, sugar, and a little plastic cup that fits right into a dolphin's blowhole" ("Americans Having Less Sex, Avoiding Movie Theaters As U.S. Braces For Possible Fifth Wave" https://youtu.be/moJJuJd5S4Q). The visceral side of the message, which stimulates a strong response in the studio audience, humanizes the message and calls for political accountability as well as a re-evaluation of individual habits, and the uncertain emotion of shame strengthens the impact of the message (Cf. Strick, 2021; Zekavat & Scheel, 2023).

The political determinants of mitigating climate change resurface in another segment, when Colbert lashes out at world leaders and their inaction before COP26. "World leaders are talking about maybe doing something about it [i.e., climate change] at the UN climate conference in Glasgow. Wait a second! Doing something about climate at a climate conference!? What are the odds? Not good!" ("Queen Urges Leaders To Rise Above Climate Politics|Preparing Kids For The Vaccine" https://youtu.be/meUe2Vg4pOo). The focus is unmistakably on how the governments involved have failed to regulate the political and economic determinants of environmental behavior amongst different sectors including corporations, organizations, and citizens. He maintains that COP26 is called so "in honor of the 26 years they've copped out of doing anything about climate change, and the stakes are super high for the planet." Likewise, Colbert also criticizes the insufficient response of G20 leaders to climate change by the following caustic comparison: "calling 911 and having the operator tell you the EMTs will be there sometime between the hour of noon and the funeral." Colbert uses satire not only to convey the urgency of the situation but also to censure the inaction of policymakers by underscoring the disparity of how things are and how they ought to be ("World Leaders Try To Wish Away Climate Change, Bezos Skips Halloween To Party With Bill Gates" https://youtu.be/-b0G_CCaTI4).

In addition to general environmental issues, the political and economic interventions of the show also take specific policies and problems on board. In one segment, for instance, it particularly targets the policy that prioritizes funds to foster resilience against climate change over financing preventive measures in the first place as stipulated in the US$1.2 trillion infrastructure bill. Colbert quips, "this is the equivalent of showing up to a car crash and handing out seatbelts" ("British Royals Can't Stop Talking About Biden's Alleged Gas Leak|Pfizer May Have A Covid Cure" https://youtu.be/MkopjQw0VBs). So the show actively promotes mitigating measures over adaptation and resilience against climate change in a specific piece of legislation. Instead of criticizing a general political concern like the inaction of world leaders, therefore, the show targets a very specific part of an unprecedented piece of legislation that will have strong consequences in that it sets the agenda for climate policies for the future. In another episode, Colbert presents data on the role of corporations and petrochemical companies in global carbon emissions, before sarcastically remarking,

> Oh! Who would have thought that oil companies would be bad for the environment, after all they've done to barbeque the Gulf of Mexico? [a picture of the boiling Gulf is depicted on the screen] Well, I say it's time to stop. Those companies ought to be ashamed of themselves, unless they're a sponsor, in which case, knock it off maybe or maybe not. Thanks for the commercials! It's how I pay for my Tesla.

This is an instance of reflexive satire (Zekavat, 2019b) that targets the host in foregrounding the economic determinants of behavior for both corporations and the media. Colbert declares the show's conflict of interest in that it chastises the corporations that are financially supporting its own existence. In other words, the show undermines the media coverage that climate change receives and implies that economic interests could explain why the narrative spun around facts is not always consistent. Having fustigated hypocrisy and tokenism, he goes on to mock himself by saying, "I'm a huge hypocrite. I'll never do anything that's inconvenient to me. That's why there has to be systemic change, backed up by government action to make everyone make the right choices, not easy ones. If it weren't for the government intervention, we'd still be eating lead paint chips, dipped in a fresh asbestos hummus, and washing it down with cocaine Colas." He targets himself to draw on a larger point: vested interest hinders meaningful environmental action, so policymakers must intervene to regulate corporate behavior. Citing instances that governmental interventions have brought concrete positive change, Colbert advocates that policymakers and governments, i.e., regulatory systems and mechanisms, must put effective policies and behavioral controls in place in order to make sure that those who benefit from environmental degradation are made to conform to sustainable policies. That being said, he takes caution not to encourage individuals to shift responsibility. He insists that the "suicidal world-destroying system of greed" of corporations does not exclude individual responsibility and action as he invites Mother Earth to inform the audience about what they can do on the individual level ("Is Systemic Climate Action Possible? Stephen Seek Guidance From Mama Earth" https://youtu.be/rbzMvJf1Vxg).

*The Late Show with Stephen Colbert*, as we have seen, manipulates both individual and social determinants of environmental behavior to encourage behavioral change among citizens, policymakers, and corporations (Cf. Baym & Shah, 2011). This, of course, does not mean that these interventions were necessarily effective. Investigating the extent to which such interventions are effective over time requires a different type of study. Arguably, one shortcoming of *The Late Show with Stephen Colbert* is that it does not aim to prepare and present a strong and focused argument to its audience. Rather, the host veers into a different topic every few jokes. Like most other late-night shows, each episode of *The Late Show with Stephen Colbert* features several short segments that do not necessarily cohere around a single topic. Even the monologue almost never concerns a single topic. Each section consists of many jokes that are thrown at the audience one after another. This makes it very difficult, if possible at all, to deliver a sustained and in-depth analysis. Besides, the show has to cut to commercials after each segment which further disrupts the flow and disturbs concentration. An alternative format of late-night shows that is not as strictly committed to infotainment and satiric news might provide greater opportunities for developing a more coherent and stronger argument. These shows do not air every night, but rather one episode per week, and they are usually shorter in length and border on investigative journalism. These features afford considerably more time and space for fashioning a better researched, more sharply focused, and more meticulously crafted message. This format, as a result, can facilitate a more sustained and cohesive treatment. In the following section, I will investigate one episode of *Last Week Tonight with John Oliver* (HBO, the US) and *ZDF Magazin Royale* (ZDG, Germany) as two examples of the latter show format.

## *Last Week Tonight with John Oliver*

A US-American late-night show with international reach, *Last Week Tonight with John Oliver* (*LWT*) premiered in April 2014 on HBO. Currently in its twelfth season, *LWT* reflects on current events and the developments of the week in rather short segments, while the main story in each episode is a more extended treatment of a single topic. The length of episodes is around 30–45 minutes, and the show touches upon a wide range of topics. Episodes are usually made available on YouTube for free which gives the show even a larger audience.

In an episode on carbon offsets, John Oliver addresses individuals and demands action from governments and corporations. The episode explains how structural flaws have given rise to empty promises of carbon offsets on which corporations are quick to capitalize. Not only do carbon offsets fail to mitigate climate change, but also their false promises can very much aggravate the situation. Opening with references to popular culture, the episode quickly transitions into the urgency of climate change. It delineates the urgency of the situation by referring to scientific data and reports that serve to inform the audience and boost the credibility of the communicator. Poking fun at the idea that giant corporations are here to solve climate change, Oliver debunks the false promises of corporations in their net-zero goals. After reviewing how corporations co-opt net-zero targets for the purpose of marketing their products in their advertisements, Oliver observes that they often resort to carbon offsets as a common tactic for marketing and advertising their products. He supports this claim by citing a study that shows two thirds of corporations rely on carbon offsets rather than actually reducing emissions in their claims to carbon neutrality. This, however, does very little, if any, good: "if the idea that you can simply invest a little money and make your carbon footprint disappear sounds too good to be true, that's because it absolutely is." Despite the fact that many scientific studies have shown offsets to be unreliable, carbon offsets are "the backbones of the environmental policies of many of the biggest polluters on the planet."

Having explained the problem and laid out the significance of the topic, the episode provides a basic definition of carbon offsets to make sure that the audience can follow the argument, before explaining several ways through which carbon offsets can do actual harm rather than ameliorating the situation. The show draws on an advertisement by the BP (formerly British Petroleum) in which an old British man explains carbon offsets by using balloons. The irony is that BP is a major polluter that manipulates carbon offsets to continue capitalizing on sustainable causes. While merely trying to define the concept, Oliver manages to drag the political determinants of behavior into the equation by playing back a short clip in which Sean Hannity, the Fox News conservative anchor, explains carbon offsets to Newt Gingrich, US-American right-wing politician, in a preposterously absurd manner. In other words, satire enables the show to observe that claims to carbon neutrality that are based on carbon offsets are frequently so empty that they deconstruct themselves. Entirely relying on carbon offsets to achieve net neutrality is as absurd as the black balloons released into the air as in the case of the BP advertisement, or presenting one's wife with jewelry in return for cheating on her in Hannity's interview.

Oliver satirizes more targets including UPS, Austin's Airport, and YepYou to better convey how capitalism has subsumed the sustainability discourse and capitalized on it by manipulating carbon offsets. Attempts to squeeze financial profits from climate change extend to marketing ridiculous plans to offset one's exhalation and one's pets. Such manipulations, however, do not remain at the level of a marketing scheme with mere economic implications. Capitalizing on carbon offsets can do real environmental damage because the revenue generated by selling carbon offsets is rarely invested in ground-breaking projects and is rather invested in projects that would have been constructed anyways; in other words, they lack additionality. To illustrate this, Oliver gives the example of how, in 2020, JPMorgan Chase & Co. claimed to achieve carbon neutrality by investing in preserving the Hawk Mountain Sanctuary in Pennsylvania, an achievement that Oliver debunks because the area has been a wild bird sanctuary since the 1930s—rather than, as he sarcastically puts it, "the Hawk Mountain Chop-Chop Zone and tree Murder Playground"—and would have been protected anyways. In other words, the money that JPMorgan Chase & Co. has put into this project has not introduced any additional source of capturing $CO_2$ from the atmosphere.

Carbon offsets have several drawbacks. First, carbon credits make it possible for corporations to prevaricate. Second, some of the money collected to offset carbon emissions seems to end up benefiting billionaires rather than the environment. In other words, businesses that genuinely attempt to offset their carbon emissions do not actually reduce their emissions because they rely on carbon offsets. Meanwhile, the financial resources generated by selling carbon credits to these businesses fail to start new sustainable projects that might genuinely offset emissions. Not only does the money invested in carbon credits not effectively offset emissions, it also hinders reducing emissions by creating the illusion that they are being offset. In this way, offsets can increase rather than decrease global emissions. Ironically, corporations buy offsets to pollute more; since these offsets are not effective, they only aggravate the ecological crisis.

This is the case not only with the corporations that intend to manipulate sustainability as a relative advantage, but also with those businesses that might genuinely try to cancel out their emissions. Oliver observes, "the problem with carbon offsets is that everyone wants to believe in them: buyers want a cheap way to make a big claim, and sellers want money for doing as little as possible." As these offsets are not efficacious, the situation is being actively aggravated, and we will end up increasing emissions. This loophole is caused by structural flaws. Carbon offset registries should be neutral mediators that regulate the system, but they are not accountable to anyone and have a strong conflict of interest as they are paid by the companies that sell the offsets.

The potential negative impacts of carbon offsets are not merely limited to ecological issues. Oliver explains how companies that plant trees to offset emissions displace indigenous people from their expropriated lands and exposes the duress and threat involved in the process. In criticizing how a tree planting project by Swedish Energy Agency, largest public funder and facilitator of sustainable companies in Sweden, displaced native people in Uganda from their ancestral lands, Oliver targets the economic determinants of behavior. Moreover, this case illustrates how socio-economic status, corporeality, racial and ethnic identity, place and belonging intersect. In other words, it is not just their socio-economic status that puts native Ugandans in a great disadvantage, but also their racial identity and place of origin increase their vulnerability. The inequality inherent in the relationship between a Swedish corporate power and aboriginal Ugandans are such strong determinants of environmental behavior that they totally overshadow the individual and psychological determinants.

Oliver concludes that the benefits of offsets are widely overstated and the harm they can do is very real. "The truth is," he says, "offsets aren't the answer here. Fundamentally, we cannot offset our way out of climate change." In other words, offsets are exposed for being ineffective behavioral controls that are manipulated for economic reasons and can negatively impact the environmental behaviors of corporations and citizens alike. Some citizens are genuinely interested in doing something to help the environment and reduce their carbon footprint, and they want to believe they have done something through buying offsets. Corporations would seize the opportunity to capitalize on these good intentions. These people would end up doing nothing, or even worse doing the opposite, as the initiative of offsetting carbon emissions backfires.

This cul-de-sac, Oliver maintains, stems from ineffective political and regulatory systems as the determinants of environmental behavior: "the bottom line is that we have an offset system that places profits over science, and the rules regulating it are just far too lax." The sheer inefficacy of the system is further held to ridicule when the audience learns that Oliver has actually established his own carbon offset registry under which hoax projects are being listed. In other words, the episode clearly and concretely depicts the failure and absurdity of carbon offsets and reveals them to be a capitalist market strategy. The episode closes on Oliver's sarcastic high note of "we're making big claims while doing very little" ("Carbon Offsets: Last Week Tonight with John Oliver [HBO]" https://youtu.be/6p8zAbFKpW0).

We have seen that the episode primarily concerns the social determinants of collective environmental behavior in targeting large corporations and structural loopholes, yet Oliver does not stop short of reaching out to individuals chiefly by referencing popular culture. The show does not ignore its own status as a popular cultural production and the workings of the culture industry. The host seeks to reach out to individuals by referencing popular culture; for instance, he pokes fun at some bands that have pledged to net-zero targets through carbon offsets. Alongside his attacks on institutions and corporations, several humorous references to popular cultural productions and Oliver's impressions of popular culture characters can be seen as attempts to appeal to individuals. Previous research has conveyed that *LWT* has been successful in mobilizing its audience in achieving socio-political impact. In their study of the impact of satire for garnering activism toward Net neutrality, for instance, Meier and Berg (2022) conclude that Oliver's satire can draw attention, captivate the audience, and promote activism. They argue that the "the appeal of Oliver's satirical combination of judgment and affect evidences satire's capacity to captivate users and inspire activism by leveraging capacities to surprise audiences and antagonists, to reveal the participants in public debates, and to level jargon-heavy, bureaucratic barriers that can prevent would-be activists from joining the fray" (p. 16). "Carbon Offsets: Last Week Tonight with John Oliver" mostly features satirical attacks on corporations for co-opting and capitalizing on environmental concerns partly for the purpose of greenwashing; it also employs some tested strategies that had previously afforded results in socio-political interventions to persuade and galvanize its audience. In the next section, I will show how Jan Böhmermann tries to balance his call for cultural and policy change by addressing both policymakers and individuals in a similar format to that of *LWT*.

## *Das ZDF Magazin Royale*

A German weekly late-night show that similarly focuses on a single topic in each episode, *Das ZDF Magazin Royale* is a satiric program that has aired on Friday nights since 2020 on ZDF, a German public service television. Each episode is around 30 minutes long. Jan Böhmermann, the host, has been a controversial media figure in Germany and in Europe. His satiric attack against the Turkish President Erdoğan in 2016 provoked a strong reaction on the part of the Turkish government and tested and pushed the limits of satire under German penal code. The format of the show, its publicity campaigns, and parrhesiastic tone are similar to *Last Week Tonight with John Oliver*. While the show has been running for a relatively shorter time, it has dedicated several episodes to environmental issues, particularly addressing their socio-political context.

In one episode on "Jetzt schon Wahlsieger: das Auto!," for instance, Böhmermann satirizes how politics and politicians have served the automotive industry in Germany, while sacrificing the rights of citizens, particularly children. Car culture is very prominent in Germany partly because the contributions of car manufacturers to the country's GDP and economic growth are considerable, to say the least. The episode opens with the host praising a car on exhibit in the studio. Böhmermann claims to own the car and asks, "does anyone with a car like this need any personality?"[6] This is a sarcastic remark on the national car culture in which people try to compensate for what they lack through exhibiting their cars. More specifically, he quips, "the larger, the faster, and the more expensive the car, the better the individual who is driving it." The sarcasm again attacks people who think they can compensate for voids in their personality by exhibiting their material belongings. It seems that Böhmermann strives to demystify the false assumption correlating wealth

[6] All translations are mine.

with personality in the first step and calls for changing attitudes, beliefs, values, and norms at a collective level.

The host continues to expose and chastise such harmful attitudes, beliefs, and values that shape Germany as a country made for cars rather than for people: "Auto first, Mensch second." Cars first, and only then people. Cars are better than people because other people can break your heart, but your car cannot. That's the reason why the automotive industry controls people, children, and politicians in Germany. Maintaining the satiric tone, he adds, "car stands for freedom. The relationship between car and freedom is similar to the relationship between car and Christian Lindler." Christian Lindler, the chairman of FDP, the Free Democratic Party, represents a liberal voice that advocates economic liberalism, free market, and reducing governmental interventions to the minimum, and insists that laws and regulations should not be passed unless they are absolutely necessary. This is particularly pertinent as the episode was aired at a time when introducing a national speed limit proved to be controversial before the general elections in 2021.

Böhmermann is quick to poke fun at Lindler's hypocritical ignorance and continues to reprimand attempts to connect cars with protecting freedom, common sense, and progress. A montage of the politician's interviews is played where he states that drivers are paying too much tax in Germany. Lindler claims to be as liberal as Montesquieu and against legislating unless it is absolutely necessary to which Böhmermann responds with a series of nonsensical sounds: "Ja! Mon Dieu! Montesquieu! Holleri! Du Dödel die die Dödel dö!" Lindler seems to refer to Montesquieu when he writes, "all punishment which is not derived from necessity is tyrannical. The law is not a mere act of power; things in their own nature indifferent are not within its province." But the fact that Lindler rips this idea from its context raises doubt if he is simply manipulating ignorant pretensions to liberalism to push forward his political agenda. What is actually meant by Montesquieu is that "liberty … requires that the laws concern only threats to public order and security, since such laws will protect us from harm while leaving us free to do as many other things as possible. Thus, for instance, the laws should not concern offenses against God, since He does not require their protection" (Bok, 2018).

So Böhmermann starts at a personal level but soon shifts his focus to the cultural and political determinants of behavior before connecting them to larger political considerations. In other words, he refuses to see car culture as simply a personal preference but rather acknowledges how personal behavior is enmeshed in cultural, social, and political spheres. Employing reflexive satire, the host ironically adopts a persona with a mindset he is satirizing. Calling cars the last barricade against common sense, he maintains that cars stand for "Freiheit, Fortschritt, Festellbremse." The bathos from freedom and progress to handbrake is complemented by a picture of a huge truck on the back of which is written an insult to Greta Thunberg. Sustaining the irony, the host announces that the episode concerns cars and their important role in Germany. He goes on to indignantly ask why political candidates receive the delegates of the automotive industry in outlandish feasts before the elections, while the police arrest environmental activists demonstrating against cars, and the media calls them idiots.

The host shows that some of the solutions proposed by policymakers do not genuinely engage the problem. Carbon capture, for instance, is presented as if it were a viable solution to tackle climate change while there is real harm in promoting impractical solutions to capture carbon instead of cutting down on the number of cars to save energy and reducing emissions. Böhmermann's criticism simultaneously reflects on the national obsession with cars, an obsession that has blinded people to the scale of the problem and its viable solutions. He concludes by calling attention to the deploringly absurd situation that "our cities are built for cars rather than for people."

The episode pays equal attention to cultural determinants of environmental behavior alongside political factors. In discussing an instance of popular culture in a Disney Pixar movie, Böhmermann expresses his concern over the socialization and education of children, a prominent form of ideological state apparatuses. Education is just one of the ways through which attitudes, beliefs, and norms are instilled in children. But education is not alone in potentially fostering environmentally destructive behaviors; the use of tropes in advertisements is yet another example of such manipulation. The episode reviews several car advertisements in order to convey how they represent different ecosystems as "natural" habitats of cars. A montage of different advertisements depicts a sedan breaking hard in the woods to avoid crashing deer, a jeep driving in a river, a station wagon parked in the forest, a hatchback parked by a pond with animals around it, cars in deserts and steppes beside animals. They invariably depict nature as the natural habitat of cars. This makes Böhmermann conclude that, "Autos sind praktisch Natur" or cars are practically [part of] nature. In this way, the show manages to illustrate what Greg Garrard had argued in the first edition of his *Ecocriticism* back in 2004: Automotive and advertisement industries co-opt a metaphor to create, present, and perpetuate a false impression about the relationship between cars and nature (see also Garrard, 2023). He is quick to expose this false impression clearly as an instance of greenwashing by the automotive industry. Corporations manipulate the metaphoric juxtaposition of cars and nature to make false or misleading impressions and claims about the environmental affinity of their products to nature.

The episode does not stop at merely making claims but provides evidence and examples to illustrate them. In order to counter-argue the false impressions created by advertisements, the host strives to disillusion his audience by talking to an expert who emphasizes that even hybrid cars have a carbon impact both during the production process and when being driven, thus enhancing his credibility (Cf. Transport & Environment, 2025). He also notices that cars have been growing in average size over the past decades. Besides the negative environmental impact of larger cars, they can also pose serious hazards for pedestrians, and in particular children. This abstract idea is concretely illustrated by an incongruous picture of a child in front of a truck, showing he cannot reach higher than the front grill.

All these culminate in the host asking his audience whether they should pose a few uncomfortable questions to politicians before the election. He is aware that the public debate over cars, and particularly a national speed limit, can sway the voters' opinions. At the same time, he is also aware that he does not need to appeal to more progressive voters as it would be preaching to the choir. So instead of adapting an environmentalist persona, he poses in mid-center only by distancing himself from the autobahn mania. Nor does he build his entire argument around the environmental impacts of cars but rather on their impact on our well-being, especially in the case of children. Using these persuasion strategies, he tries to appeal to the widest target audience across the spectrum of political persuasions and beliefs.

Böhmermann proposes that riding bicycles could be a solution to the problems that cars pose in cities while noting that the infrastructure is not yet ready. He is aware that the tradition of car culture as a motivational type might not readily activate personal pro-environmental norms (de Groot & Thøgersen, 2018); therefore, he strives to present the tradition in a different light. Acknowledging that Germany has a long pedigree in the automotive industry, he goes even further back to 1817 when Carl von Drais invented the bicycle in Mannheim, Germany. Another video dating back to 1902 records a very different urban environment on the streets of the country. Deviating from this pre-war pedestrian-orientated development, post-war Germany rebuilt her cities for cars not for people in the wake of the World War II destruction. What was not destroyed in the war was demolished to make room for wider streets and more parking spaces.

The increasing number of cars on the streets ironically gave rise to another problem in the country: children! They were considered to be an impediment to the flow of traffic, and playgrounds were Germany's solution to this new urban problem. Authorities put children in playgrounds so that cars can freely use the rest of the city. While the host notes that all playgrounds seem to be replete with car parts, we see several pictures of playgrounds on the screen where swings, seesaws, and sand pits are made of car tires. Reacting to an old PSA in which a car crashes into a dummy child, Böhmermann exclaims, "Oh! No!" before consoling his audience by saying, "No worries. It wasn't a real car." While the audience worries about a child being hurt, his persona is ironically concerned for the car in a city that prioritizes cars over children. Constant exposure to such a priority over decades has resulted in the public's misconception that children rather than cars are a hazard for traffic and drivers. Akin to exposing the workings of a metaphor in the advertisements that are set in nature, satire allows Böhmermann to expose another fallacy driven by the use of public rhetoric. Böhmermann employs satire, positions himself in center politics, appeals to traditional and national values, and foregrounds the safety and wellbeing of children to reach out to more conservative voters that might not place environmental concerns as their top priority.

Böhmermann warns that the auto-culture is so ingrained in Germany that the country seems willing to descend to what he dubs communist authoritarianism to uphold it. It is not just the rights of children but also those of adult citizens that are scarified for the car culture as many people have lost their homes to make room for new roads. He seeks the reason behind such unfair infringements in the lobbying power of the automotive industry in German politics. He supports his claim by providing several examples to show how politicians and top figures in the automotive industry frequently exchange roles—or what he calls "incest between politics and the automotive industry"—to further intertwine these two fields of interest and power together. These include Hildegard Müller, the Minister of State to the Chancellor of the Federal Republic of Germany (2005–2008) and President of the most prominent lobby of the automotive industry, German Association of the Automotive Industry since 2020; Eckart von Klaeden, the Minister of State at the German Chancellery (2009–2013) and now Daimler's chief lobbyist; and Joachim Koschnicke Opel's chief lobbyist (2013–2017) and then directly afterwards CDU's[7] campaign manager. Such potential conflicts of interest can explain the overblown power of the automotive industry in German politics and political decisions. Political power and influence inevitably serve economic interests at the cost of the larger public and global good. Using satire to frame the situation, Böhmermann laments that little is likely to change after the election, as parties across the political spectrum—including those that define themselves by their environmental commitments—fail to recognize the urgency of the climate crisis and have put forward no concrete plans to address it. In the end, the host demands that streets should be reclaimed by people before recapitulating what he has discussed in this episode in a satiric song ("Jetzt schon Wahlsieger: das Auto!|ZDF Magazin Royale" https://youtu.be/ybTHWzmlw70).

## Conclusion

I tried to explore the use of humor and satire for the purpose of environmental advocacy in three late-night shows. By analyzing *The Late Show with Stephen Colbert*, *Last Week Tonight with John Oliver*, and *ZDF Magazin Royale*, I conveyed how humor and satire can be employed to address the individual and social determinants of behavior among citizens, corporations, and politicians. All three shows acknowledge and address both the individual and social (cultural, political, and economic) determinants of environmental behavior. While calling for personal responsibility at the individual level, they are cautious not to overcharge individuals in tackling a complex and multifaceted issue like climate change. Instead, they demand collective action in introducing more effective policies and criticize corporations and their political enablers for their irresponsibility.

While they employ humor and satire to respond to political and environmental concerns, these late-night shows tend to muster specific strategies for addressing these topics at individual and collective levels. They tend to use lighter comic styles including humor when trying to reach out to and persuade individuals to reconsider their environmental behavior

[7] The Christian Democratic Union of Germany, a prominent political party in Germany.

and spare darker styles including satire and sarcasm for attacking politicians and corporate greed (Cf. Ruch et al., 2018 for comic styles). In this way, they attempt to strike a balance between creating positive emotions in their audience and tempering their expectations for such emotions. Besides availing themselves to different comic styles to address different stakeholders, these shows are also aware that being comics might undermine their credibility, so they frequently cite academic studies or interview scientists to mitigate this drawback.

The main difference in how these shows use humor and satire for environmental advocacy is probably due to their formats. The format used by *Last Week Tonight with John Oliver* and *ZDF Magazin Royale* allows them to focus on and delve into specific topics and provide more comprehensive critiques. *The Late Show with Stephen Colbert*, by contrast, covers a larger diversity of topics but does not put forth a well-researched and comprehensively developed critique. Another interesting disparity is that both Colbert and Böhmermann adopt personas and employ reflexive humor and satire to communicate environmental messages. In other words, they do not spare themselves as the object of satire. Oliver does not use reflexive humor and satire in this episode, although he is quite fond of them in his work. This is probably because he is trying to expose the inefficacy of a supposed solution to climate change and needs optimal credibility to drive his point home. As using reflexive humor and satire can undermine his credibility, he decides to keep them at bay.

I tried to show that late-night satire can provide a viable strategy for environmental communication and advocacy. This, of course, does not imply that satire is a panacea (Cf. Zekavat, 2023). Holistic models for environmental behavior and a more nuanced approach to the subtleties of humor and satire, however, can guide public communicators in forming more effective messages and map out pertinent intricacies in designing research for academics.

The analysis of late-night shows further reveals a preference for humor over satire, with the latter being more resistant to easy narratives and willing to engage with ambivalent emotions. Satire does not necessarily promote a narrative of hope or positive emotions but instead creates space for undefeated despair that can potentially empower the Exhausted of the Earth, emphasizing complex material entanglements and the notion of living in end times. While acerbic satire could potentially serve as a more effective strategy for fostering critical engagement, late-night shows often shy away from it due to their operational focus on entertaining audiences.

These shows partly function as a form of escapism rather than a platform for serious critique, as excessive cognitive and emotional demands might threaten their business viability.

The analysis also shed light on a certain level of awareness among these shows regarding psychological mechanisms and behavioral determinants. By incorporating psychological insights into their narratives and often drawing on popular culture, these shows eventually reinforce the culture industry's dominance. This duality offers an opportunity to understand how the culture industry insidiously leverages psychology to optimize its influence, while simultaneously opening up spaces for resistance and subversion (Giamario, 2023). By decoding these psychological mechanisms, viewers can potentially appropriate them for alternative purposes.

Despite this, the primary aim of late-night shows remains appealing to audiences and prioritizing profit. Consequently, they are less inclined to promote undefeated despair or endorse radical approaches such as grassroots organizing or systemic refusal. Their perspective tends to reflect a Western, predominantly male, middle-class, white, industrialized, and neoliberal democratic viewpoint, often neglecting the differentiated impacts of ecological mutations in various regions, particularly the Global South. Ultimately, their complicitous critiques are framed within the confines of neoliberalism and its institutions, favoring incremental reform over disruptive change. In the words of Hardt and Negri (2000, p. 33), "the legitimation of the imperial machine is born at least in part of the communications industries, that is, of the transformation of the new mode of production into a machine." However, by occasionally exposing the fissures of the inner workings of neoliberalism and consumer capitalism, these shows may inadvertently offer opportunities to challenge or defy these systems from within.

## References

Apter, M. J. (2013). Developing reversal theory: Some suggestions for future research. *Journal of Motivation, Emotion, and Personality, 1*(1), 1–8. https://doi.org/10.12689/jmep.2013.101

Baym, G., & Shah, C. (2011). Circulating struggle: The online flow of environmental advocacy clips from the daily show and the Colbert report. *Information, Communication & Society, 14*(7), 1017–1038. https://doi.org/10.1080/1369118X.2011.554573

Bok, H. (2018). Baron de Montesquieu, Charles-Louis de Secondat. In E. N. Zalta (Ed.), *The Stanford encyclopedia of philosophy*. Plato. https://plato.stanford.edu/archives/win2018/entries/montesquieu/

Bore, I.-L., & Reid, G. (2014). Laughing in the face of climate change? Satire as a device for engaging audiences in public debate. *Science Communication, 36*(4), 454–478. https://doi.org/10.1177/1075547014534076

Chattoo, C. B., & Feldman, L. (2020). *A comedian and an activist walk into a bar: The serious role of comedy in social justice*. University of California Press.

De Groot, J. I. M., & Thøgersen, J. (2018). Values and pro-environmental behaviour. In L. Steg & J. de Groot (Eds.), *Environmental psychology* (pp. 167–178). Blackwell. https://doi.org/10.1002/9781119241072.ch17

Garrard, G. (2023). *Ecocriticism* (3rd ed.). Routledge.

Giamario, P. (2023). *Laughter as Politics: Critical Theory in an Age of Hilarity*. Edinburgh University Press.

Hardt, M., & Negri, A. (2000). *Empire*. Harvard University Press.

Hendrix, C. S., Koubi, V., Selby, J., et al. (2023). Climate change and conflict. *Nature Reviews Earth and Environment, 4*, 144–148. https://doi.org/10.1038/s43017-022-00382-w

Hutcheon, L. (1988). *A poetics of postmodernism: History, theory, fiction*. Routledge.

Meier, M. R., & Berg, S. V. L. (2022). The activist potential of networked satire: Last week tonight with John Oliver and the struggle for net neutrality. *Western Journal of Communication, 88*(1), 3–32. https://doi.org/10.1080/10570314.2022.2130003

Milman, O. (2021). Climate denial is waning on the right. What's replacing it might be just as scary. *The Guardian*. https://www.theguardian.com/environment/2021/nov/21/climate-denial-far-right-immigration

Norgaard, K. M. (2011). *Living in denial: Climate change, emotions, and everyday life*. MIT Press.

Panda, P. (2020). Indigenous humor in Thomas King's the back of the turtle: An ecocritical perspective. *Studies in American Humor, 6*(2), 323–340. https://doi.org/10.5325/studamerhumor.6.2.0323

Remnick, D. (2019). Alexandria Ocasio-Cortez is coming for your hamburgers! *The New Yorker*. https://www.newyorker.com/news/daily-comment/alexandria-ocasio-cortez-is-coming-for-your-hamburgers

Ruch, W., Heintz, S., Platt, T., Wagner, L., & Proyer, R. T. (2018). Broadening humor: Comic styles differentially tap into temperament, character, and ability. *Frontiers in Psychology, 9*, 6. https://doi.org/10.3389/fpsyg.2018.00006

Seymour, N. (2018). *Bad environmentalism: Irony and irreverence in the ecological age*. University of Minnesota Press.

Shome, D., & Marx, S. (2009). *The psychology of climate change communication: A guide for scientists, journalists, educators, political aides, and the interested public*. Center for Research on Environmental Decisions.

Skurka, C., Niederdeppe, J., & Nabi, R. (2019). Kimmel in climate: Disentangling the emotional ingredients of satirical monologue. *Science Communication, 41*(4), 394–421. https://doi.org/10.1177/1075547019853837

Skurka, C., Niederdeppe, J., Romero-Canyas, R., & Acup, D. (2018). Pathways of influence in emotional appeals: Benefits and tradeoffs of using fear or humor to promote climate change-related intentions and risk perceptions. *Journal of Communication, 68*(1), 169–193. https://doi.org/10.1093/joc/jqx008

Strick, M. (2021). How humor can promote central-route persuasion: The role of ambivalence. In M. Strick & T. E. Ford (Eds.), *The social psychology of humor* (pp. 3–19). Routledge.

Transport & Environment. (2025, October). Smoke screen: the growing PHEV emissions scandal — Long-range PHEVs and EREVs are a diversion on the road to zero emissions. Authors: Sofía Navas Gohlke, Yoann Gimbert. Expert group: Lucien Mathieu, Richard Riley. Editor responsible: William Todts. Brussels: European Federation for Transport and Environment AISBL. Retrieved from https://www.transportenvironment.org/uploads/files/2025_10_PHEV_smoke_screen_report.pdf

Walker, K. (2021). As climate displacement increases, migration myths fuel fears. *Equal Times*. https://www.equaltimes.org/as-climate-displacement-increases

Wyer, R. S., & Collins, J. E. (1992). A theory of humor elicitation. *Psychological Review, 99*(4), 663–688. https://doi.org/10.1037/0033-295X.99.4.663

Yeo, S. K., Becker, A. B., Cacciatore, M. A., Anderson, A. A., & Patel, K. (2022). Humor can increase perceived communicator effectiveness regardless of race, gender, and expertise—If you are funny enough. *Science Communication, 44*(5), 593–620. https://doi.org/10.1177/10755470221132278

Zekavat, M. (2019a). Reflexive humour and satire: A critical review. *European Journal of Humour Research, 7*(4), 125–136. https://doi.org/10.7592/EJHR2019.7.4.zekavat

Zekavat, M. (2019b). Satire, humor and ecological thought. *Neohelicon, 46*(3), 369–386.

Zekavat, M. (2023). The ambivalent affordances of humour in capitalist organisations. *European Journal of Humour Research, 11*(1), 184–200. https://doi.org/10.7592/EJHR2023.11.1.728

Zekavat, M., & Scheel, T. (2023). *Satire, humor, and environmental crises* (1st ed.). Routledge.

CHAPTER 5

# Authorized by the Department for Going Gentle into that Good Night

In the previous two chapters, we examined how operating within the constraints of the culture industry imposes certain constraints and can lead to the appropriation and co-optation of the narrative, thereby limiting the transformative potential of satire. As Coulson (2007, p. 144) notes, "the culture industry forces the masses to desire their own repression and reification through its mimesis of a subject and object reconciled in undominated nature." To better appreciate how the dynamics of the culture industry impacts vibrant materiality, it is imperative to emphasize that for Horkheimer and Adorno, "the original separation of subject and object—the opening of the dialectic and of domination—comes from the cry of terror at the sight of undominated nature" (Coulson, 2007, p. 144). In other words, the metabolic rift can explain anthropocentric attempts to dominate not only nature but also social subjects. Both forms of domination seem to originate in a desire to subjugate the vibrancy of matter.

This constraint is partly attributed to the economic and political mechanisms and interests that shape the operations of the culture industry. While decentralized social media platforms appear to offer new spaces for expression, they are not entirely free from conflicts of interest, such as those associated with state funding or reliance on sponsor and advertising revenue. Although these creators can rely on patrons and revenue from platforms like YouTube, they keep operating within a framework of platform capitalism (Srnicek 2017) and techno-feudalism (Varoufakis, 2023).



M. Zekavat, *Leveraging Satire for Environmental Advocacy*,
https://doi.org/10.1007/978-3-032-04838-7_5

This chapter explores how the *Honest Government Ads* (HGA) series wields the affective and performative affordances of satire as a tool for socio-political intervention, with a particular focus on environmental advocacy (Bogad, 2016). Produced by Juice Media in Australia, HGA is a collection of satirical advertisements that parody official public service announcements (PSAs). Juice Media was launched on YouTube in 2008, with the inaugural episode of Honest Government Ads released on May 28, 2016. Through its productions, HGA addresses a range of issues, including colonialism and various social, economic, and political topics, such as elections and environmental concerns. By employing satire, the HGA series seeks to engage with political discourse and sway public opinion, using satire to critically examine and challenge prevailing narratives.

Examining the impact of cultural productions from Australia is particularly pertinent in the current context, in which efforts to combat climate change are (perceived as) inadequate and progressing too slowly, despite growing global concerns over climate urgency. Australia stands out as a major producer and exporter of gas and coal, and when considering its relatively small population, it ranks remarkably high in per capita production. The fossil fuel and mining industries have significantly impacted the country, contributing to environmental issues such as the bleaching of the Great Barrier Reef, the world's largest coral reef system, and recurring widespread wildfires. Moreover, Australia's considerable emissions are not solely due to its fossil fuel and mining activities but are also linked to farming and high meat consumption. Despite not being recognized as an example of environmental leadership, Australia is set to co-host the 31st Conference of the Parties (COP31) in 2026. In this context, exploring the potential of satire for environmental advocacy becomes increasingly significant, as it offers a means to engage with and critique these pressing issues in anticipation of the upcoming international climate conference.

Although Juice Media employs satire to engage individuals, its primary targets are corporations and policymakers, with the aim of influencing public opinion to effect changes in political decisions and corporate behaviors through voting. This chapter will analyze episodes of *Honest Government Ads* that focus on critical environmentalism to elucidate how they aim to optimize the impact of their environmental communication at both individual and collective levels. HGA strategically eschews positive emotions, choosing instead to highlight uncertainty emotions. It further foregrounds intersectionality by addressing ecological issues in conjunction with social justice, thus tackling complex problems in a holistic manner. Through this nuanced use of satire, HGA seeks not only to inform and engage viewers but also to inspire thoughtful and informed action on environmental issues.

## The Rich World

While predominantly focused on Australian issues, HGA series also offers keen insights into environmental issues globally. One notable example is its critique of the US Supreme Court's decision to limit the authority of the US Environmental Protection Agency (EPA) in regulating emissions. In this episode, the narrator underscores the expansive power and authority of the court, equating it with the US government itself. The episode raises concerns about the court's inclination to wield its significant power over partisan issues, particularly following recent appointments by President Trump, which has shifted the partisan balance to a 6–3 majority favoring Republicans. The narrator comments, "now that we have enough Federalist Society nutjobs on the bench, we'll be using this Court's unchecked power to wrong some rights … like … the right for polluters to kill even more of you in heatwaves, floods, and fires" (Juice Media, 2022c). This sharp satire highlights the potential consequences of a Repressive State Apparatus, namely judicial power, on environmental policy and behavior. While the episode critiques the power dynamics and unchecked authority that appear to conflict with the principles of elective democracy, it ultimately encourages the audience to participate in the electoral process by voting in the upcoming midterms. This suggests that, despite its critical stance, it does not envision a political space beyond the existing framework of two-party representative democracy in the US. By advocating for voter participation within this established system, HGA implicitly acknowledges the limitations of the current political landscape while still emphasizing the importance of civic engagement as a means of influencing policy and holding power to account.

Environmental concerns are also prominently featured in HGA's critique of the British Tory Government, particularly in its handling of wildfires, sewage issues, dying rivers, drought, and crop failures. One episode highlights past policy failures, such as former Prime Minister David

Cameron's (PM 2010–2016) rollback of weatherproofing and renewable energy schemes in response to the 2013 energy crisis. The "cutting the green crap" policy is noted for adding billions to energy bills, pointing to short-sighted decisions with long-term consequences. Similarly, the episode criticizes Prime Minister Liz Truss (PM 2022) for dismantling green levies and renewable initiatives in response to the energy crisis exacerbated by the Russian invasion of Ukraine, illustrating a failure to learn from past mistakes and increasing dependence on gas, the root cause of the ongoing crisis.

HGA not only critiques these governmental actions but also offers solutions, such as taxing companies like Shell and BP and re-nationalizing more energy companies. The series portrays the government's neglectful attitude with the statement, "British Government. You're on your own, dickheads," and suggests civil disobedience, including strikes and unionizing, as means to address the leadership void. Using irony, the episode encourages its audience to join the "Don't Pay" campaign, which threatens to cancel direct debits if energy prices continue to soar, reminding viewers of the historical success of similar movements, such as the "We Won't Pay" campaign that contributed to Margaret Thatcher's downfall in 1990 (Juice Media, 2022b).

Through these depictions, HGA clearly targets the government, aiming to provoke political action rather than individual behavioral change. It aligns itself with the public against policymakers and corporations, advocating for systemic change through collective action. By emphasizing resistance methods like unionizing, striking, and taxing the wealthy, the series underscores strategies that prioritize environmental interests for the Exhausted of the Earth. In essence, HGA advocates for externalization and empowerment of the Exhausted, encouraging them to exert pressure on governments and corporations to enact meaningful change.

Amidst numerous allegations of corruption and incompetence, HGA focused on the critical state of the Amazon rainforest as Brazil approached its presidential election in 2022. The ad enumerates the environmental catastrophes exacerbated by President Bolsonaro's administration, particularly emphasizing the dramatic increase in deforestation and fires, which have pushed the Amazon to the brink of ecological collapse. The episode explains how the rainforest, once a major carbon sink, has become a net emitter due to policies that undermine environmental agencies in favor of illegal mining and logging, and allow the agricultural sector to clear vast areas for soy and beef farming.

HGA takes a holistic approach to environmental mutations, highlighting that these issues are interconnected with social and economic inequalities. It critiques the export-driven agricultural practices that prioritize profit over addressing hunger and poverty in Brazil, with a scathing remark: "Hungry Brazilians can go eat a bag of dicks. AGRO is Tree stumps. AGRO is Starving population. AGRO is shiftfuckery." Furthermore, the ad condemns the government's tacit approval of landgrabbers who endanger indigenous communities, emphasizing the symbiotic relationship between these communities and the rainforest. This underscores the far-reaching implications of misguided policies, threatening not just natural resources but also the lives, homes, and cultures of those who have resided in these forests for centuries.

Rather than simplifying the issue by urging individuals to make minor lifestyle changes, HGA reflects on the Brazilian government's inadequate response to these environmental disasters, highlighting the disparity between intentions and actions at the collective level. It criticizes the government's denial of the fires and the dismissal of the head of the science agency following a dispute with Bolsonaro over deforestation data. Acknowledging that the general audience may not grasp the scientific complexities of environmental crises, HGA uses satire and humor to make these concepts more accessible. For instance, it uses the term "cubic fuckton of carbon" instead of "123 gigatons of stored $CO_2$" to help audiences understand the magnitude of the issue, with the presenter in a lab coat using exaggerated gestures to enhance the comedic effect. Humor serves to make complex topics more engaging and memorable while underscoring the incongruity of the situation.

The episode also highlights the political and economic determinants of environmental behavior, linking these factors to the upcoming presidential election, in which Bolsonaro sought a second term. HGA accuses him of adopting a strategy similar to Trump's by questioning the integrity of the election results. The ad satirically suggests, "and if that happens [i.e., if we lose the election] we hope our imperial masters can come and help us end democracy here just like they did all those times in Latin America," while displaying information about US involvement in regime changes across the region. By doing so, HGA extends the responsibility for Brazil's sociopolitical and environmental challenges to external powers, specifically highlighting the role of imperialism and US interventions in contributing to ecological degradation, dispossession, and displacement. The ad therefore comments on the intersection of environmental sustainability, economic justice, and imperialism, urging a broader understanding of these interconnected issues (Juice Media, 2022d).[1]

## Australia and Climate Change

Despite its global commitment, the primary focus of the HGA series remains to be Australia and its government,[2] where it seeks to expose and critique policy decisions and actions while providing information to citizens. Through its satirical prism, the series advocates for political action, aiming to engage its audience in a meaningful discourse on environmental and political issues. HGA does not shy away from employing negative and uncertain emotions, such as shame, in its environmental satire. By leveraging these emotions, the series seeks to provoke reflection and galvanize audiences toward advocating for change, thereby highlighting the urgent need for more responsible and sustainable governmental policies.

In one episode, the presenter impersonating the Australian Government delivers a sardonic message to the world, stating, "as cities bake, fires rage, reefs die, jet streams weaken, and six Sigma events cause climate scientists to shit their dacks, … don't despair. Australia's here to help. No, not by phasing out coal and gas; by inviting the world to a big fucking dinner party." This biting commentary criticizes the Australian Government's bid to host COP31 in 2026, describing it as an effort to refurbish the country's public image on the international stage after "a decade of climate inaction." Despite claiming commitment to climate leadership, the new government continues to approve new coal mines, fracking projects, and the expansion of gas production and carbon credits, revealing the hosting bid as a mere publicity stunt.

[1] For more examples of international environmental satire see; https://youtu.be/u7s-BgfcFXw?si=7inW1tHSpxg3L5i1

Honest Government Ad|Israel & Gaza; https://youtu.be/L0Zb9iUi0JM?si=D0nbB_ZMuJdjRKU0

Honest Government Ad|Japan v. Paul Watson https://www.youtube.com/watch?v=QqzOAyXSJMI

[2] Honest Government Ad|Visit Australia!; https://youtu.be/V41-uYPXw6I

Honest Government Ad|Visit Australia!; (Season 2) https://youtu.be/J6vvpQEaRdk

Honest Government Ad|Visit Tasmania! https://youtu.be/EfcqCdGqjWQ?si=l7ySgN0Us-J7JrE_

Visit Tasmania|Government Approved Version; https://youtu.be/K64JNg4Nwqk?si=hjNY0F3yR_pMiP8-

HGA further underscores the hypocrisy of Australia's actions by highlighting its co-hosting bid with Pacific Island Countries, recognized as genuine climate leaders. Despite these nations' calls for Australia to cease approving new fossil fuel projects and end subsidies, the Australian Government has rejected such conditions, leveraging its aid for climate damage as a counterbalance. This aid, however, amounts to $700 million every 5 years, a stark contrast to the $11 billion in annual subsidies given to the oil and gas industries responsible for significant climate damage. The ad sarcastically notes, "but what did you expect? When we told you we'd stand shoulder to shoulder with our Pacific family, we didn't mean against climate change! We meant at the canapé table at COP31."

This satire extends beyond Australia, referencing the United Arab Emirates' hosting of COP28, where an oil company executive presided over the conference. HGA uses the "Authorised by the Department for Orchestrating our Glorious Climate Makeover and Canapé" tagline to employ the parrhesiastic power of satire to speak truth to power. It aligns with smaller, less endowed nations disproportionately threatened by rising sea levels, exposing how aid is used as a colonial tool to ensure capitalist gains at home. The series ties environmental degradation, capitalism, colonialism, and environmental justice together, critiquing how corporate economic interests overshadow the needs of people and influence political decisions and policies. In doing so, it highlights the reality that governments often serve corporate interests rather than those of the citizens who elected them (Juice Media, 2023a).

HGA raised similar concerns for COP26 in Glasgow in 2021, opening the episode with a stark message from the Australian Government to the world: "Fuck you!" This bold declaration sets the tone for the episode, which satirically forecasts that Scott Morrison, dubbed the Australian "Prime Marketer," will present hollow gestures in Glasgow. For those needing further clarification of his facade, the episode quips, "for those of you not fluent in Cuntonese [mind the pun on Canberra/cunt], the official language of the Australian Government, what he is saying is: Fuck you! We'll keep digging, burning, and exporting fossil fuels, let you do the hard work, and then take all the credit." Despite Morrison's rhetoric, the underlying message is a steadfast refusal to change course. The federal government not only undermines states' efforts to reduce emissions but also takes credit for their accomplishments. It even claims credit for opposition policies or "things nobody does like carbon credits for not clearing land that wasn't ever gonna be cleared anyway." This occurs as the environment minister approves three new coal mines in a single month, and the minister for "emission reductions" endorses new gas projects.

The episode raises critical questions about how these policies contradict the 2050 net-zero goal, continuing to expose how the government manipulates net-zero promises to expand fossil fuel projects. The 2050 goal is non-binding, lacks policy support, and does not rely on viable technology, merely serving to "distract from what we're doing this decade." Satire again exposes the flaws of the futurity of the end of times and enables bringing us back to the end times. This critique extends beyond the Australian Government to include other nations such as Norway, Canada, the UK, and Russia, which are also failing to meet climate goals. Addressing the governments attending COP26, the presenter remarks, "we know you want to punch us in the face for obstructing every COP since Kyoto, but hey as long as you keep buying our coal and gas exports, we know you secretly love us for it." HGA not only targets fossil fuel producers but also criticizes the states that purchase and consume these resources. As long as the profit chain and fossil fuel dependency persist, there is little hope for effective and meaningful change. The episode appropriates the discourse of capitalism and familiar market language to underline that even countries not producing oil and gas—those appearing to embrace a green transition—are complicit enablers. By buying and consuming fossil fuels, they indirectly obstruct climate goals.

The episode concludes by asserting that the blame should not be placed on Australian citizens, not even on the "dickheads who keep voting for us," because, as the ad states, "for them, we reserve our biggest fuck-you of all: with our 'Technology not Taxes' plan which means you pay for bullshit technology that greenwash fossil fuel companies, and they keep paying no taxes." The episode effectively argues that the government is imposing financial burdens on its citizens—who have elected them—to benefit the fossil fuel industry, which ultimately harms both the public and future generations. By framing this as a squandering of the audience's financial resources and a breach of governmental trust, HGA raises the stakes of its argument. The series employs similar rhetorical techniques to critique carbon capture and offsets, further illustrating how governmental policies prioritize industry interests over genuine environmental progress (Juice Media, 2021a).

## Hope in Liberal Democracy: Democratic Participation as the Solution

So far we have seen that HGA generally demonstrates a keen awareness of the nuanced and intersectional complexities of ecological mutations both globally and in Australia. However, the series does more than merely diagnose these problems; it actively proposes solutions. Primarily, HGA encourages informed participation within representative democracies, aiming to empower citizens through the dissemination of information and disillusionment by satire. Furthermore, it critiques strategies of adaptation and resilience for their inadequacy in delivering timely and proportionate responses to accelerating ecological mutations, thereby advocating for more effective and participatory approaches to addressing ecological crises.

HGA has evolved its focus from primarily political topics in its first season to increasingly addressing environmental concerns over the following two seasons. Initially, the series concentrated on galvanizing individuals to vote or contact their Members of Parliament to influence decision-making and legislative processes. HGA relies on community building and fostering a sense of affiliation to motivate public engagement and participation, aiming to alter the behavior of governments and policymakers. The series harnesses the community-building function of satire to unite people, and simultaneously leverages it to effectively disseminate information. Through parrhesia, or speaking truth to power, HGA champions liberal values such as freedom and human rights, using its platform to advocate for informed and active democratic participation.

Information provision is a core tenet of HGA, as exemplified in an episode that critiques the Australian government's policy on electric vehicles (EVs). The episode highlights how policies can be manipulated through fearmongering for political gain, using satire to expose these tactics. It accuses the government of planning to "take [a] once in a lifetime opportunity and flush it down the dunny," underscoring the missed potential in failing to support electric vehicle adoption. The episode begins by establishing a common understanding of the importance of electric vehicles, emphasizing that the transition from combustion engines offers a "unique chance to rebuild our manufacturing industry, upskill our workforce, improve our air quality, whilst tackling the climate crisis and advancing human civilization." By framing the transition as a progressive endeavor with multiple benefits, the episode aims to rally public support and mobilize citizens to advocate for more forward-thinking policies.

Instead of embracing the progressive transition to EVs, the Australian Government chose to "orchestrate a fear campaign about EVs so … [the Liberal Government] can score points off Labor." The episode's presenter reminds viewers that the Liberal Government had previously claimed Labor's EV policy would "end the weekend." These misleading claims might have played a role in the Liberals' electoral victory, allowing them to introduce their Future Fuels Strategy—satirically referred to as FFS, or "For Fuck's Sake." Rather than implementing policies that would promote EV adoption, as seen with the incentives in place in many other countries, the Australian Government hesitates to electrify its own fleet—a move that could initiate a more affordable secondhand EV market. The lack of a plan to phase out new combustion engines has discouraged manufacturers from marketing their best and most affordable models or establishing new plants in Australia. As the episode sarcastically notes, "when it comes to 'Powering Choice' [the chyron reads FFS], creating jobs, and economic growth, our EV policy will ensure approximately fuckall." The ad further enumerates the benefits of transitioning to EVs, including energy independence, national security, and reduced health costs, arguing for a more committed and strategic approach to this critical shift.

It is noteworthy that episode neglects to account for the myriad environmental and socio-political challenges—such as resource extraction and dispossession—that accompany the production of lithium and batteries for electric vehicles (Riofrancos, 2025). Nor does it advocate for an alternative economic model but instead utilizes the logic of capitalism to advance its environmental agenda. By framing the transition to EVs as an opportunity for economic growth, driven by market forces and consumerism, the ad stops short of promoting a more equitable system that prioritizes multispecies and planetary well-being over economic expansion. Rather than suggesting that reducing the number of cars could enhance the health and well-being of people and other species (Cf. Chap. 4), the ad argues that increasing the number of EVs will reduce health costs. In doing so, it avoids endorsing more radical solutions and instead co-opts the prevailing discourse of liberal capitalism, using this familiar language to potentially influence decision-makers. This might be among the reasons why using a Tesla to symbolize EVs has not aged very well. The effectiveness of this strategy is open to debate, but the episode continues to employ the discourse that arises from and reinforces the status quo to deliver its critique. It quips that exhaust fumes are "linked to lower IQ, which might explain why you keep re-electing us," simultaneously appealing to citizens who vote Liberal while shaming them.

However, the drawbacks of the Future Fuels Strategy (FFS) extend beyond these issues. The episode points out that "the best part of our total lack of leadership on EVs is that it allows other dickheads to come up with dumbass policies." This criticism is directed at the Victorian Government, which plans to introduce a tax on EVs, thereby making them more expensive. The justification for this tax is that it will fund road maintenance, as EV drivers do not pay petrol taxes. However, the ad counters this argument by noting that petrol taxes are federal and are not earmarked for direct road investment. The policy is sarcastically likened to "taxing ex-smokers for no longer paying tax on tobacco." The ad concludes by urging its audience to contact their local Member of Parliament to "stop this bullshit," encouraging active civic participation to challenge ineffective policies (Juice Media, 2020a).

In yet another episode, HGA chastises the Australian Government for claiming credit where little is due regarding climate action. Again posing as an official representative, the presenter announces that the government will no longer rely on its Kyoto Carryover Credits to meet emissions targets, sarcastically suggesting that we "appreciate everyone patting us on the dick for that." The episode highlights how the government has been evading accountability, drawing a parallel to the Australian national cricket team's past cheating scandal to incite an uncomfortable emotion by underscoring the "cringe" factor of these actions. Over two decades, the government's policy at climate summits has been consistently lackluster, and the recent claim to abandon the use of Kyoto Carryover Credits is portrayed as a continuation of this inadequate approach.

The episode continues by offering a historical perspective to contextualize its critique. The Kyoto Protocol, adopted on December 11, 1997, and coming into effect on February 16, 2005, was designed to operationalize the United Nations Framework Convention on Climate Change. It committed industrialized countries and economies in transition to limit and reduce greenhouse gas emissions according to agreed individual targets. The Convention required these countries to adopt mitigation policies and report periodically (https://unfccc.int/kyoto_protocol). While most wealthy nations committed to reducing their greenhouse gas emissions, Australia distinguished itself by "bitching and whinging about how we need special treatment," demanding a higher baseline for counting emissions. Despite securing these concessions, the Australian Government ultimately refused to ratify the protocol. During COP18 in Doha in 2012, the commitments made in Kyoto were evaluated, revealing that Australia's emissions had risen by 28%. Nevertheless, due to the special treatment it had secured, the country remained below its target, highlighting the manipulative strategies employed to circumvent genuine climate responsibility.

At COP21 in Paris, it was decided that using Kyoto Credits runs counter to the spirit of the 2015 Paris Agreement. Despite this, then-Prime Minister Tony Abbott (PM 2013–2015) announced a "piss-weak" 2030 target of a 26% reduction, falling short of the 40–60% reduction recommended by experts to preserve the Great Barrier Reef. Australia insisted on retaining its "bullshit Kyoto Credits," even though carrying over credits from the Kyoto era is not legal under the Paris Accord. This prolonged inaction resulted in Australia having the highest per-capita emissions in the developed world by 2020. Yet, as the episode sarcastically notes, "don't worry, Scotty's [Morrison] working hard to find ways of marketing cringe to the world as climate action." These new approaches do not involve genuine emissions cuts or a net-zero target; instead, the government relies on the gas industry for economic recovery post-SARS-CoV-2 pandemic, merely relinquishing its Kyoto credits. The ad emphasizes, "and now that you know the story of our bullshit credits, you can fully appreciate the pure shitfuckery of saying that our contribution to global climate action will be to not use them."

The primary function of satire, therfore, is to inform the audience, ensuring they do not fall for the government's misleading claims. The episode concludes with the assertion that "what we're saying by pledging not to use our bullshit credits, is that instead of working harder, like the rest of the world, we just promise to stop cheating, which in a way, when you think about it, is still a way of using those bullshit credits. Cringe." By comparing Australia to other countries, the ad aims to mitigate the better-than-the-average effect, which is the tendency to overestimate one's pro-environmental actions. It explicitly states that Australia is not performing better than average. Furthermore, the comparison serves to foster and reinforce descriptive norms, highlighting what others actually do and what is typically expected. Through these comparisons, the ad exposes that Australia does not meet the standard, effectively using satire to critique the country's inadequate climate action (Juice Media, 2020c; see also Juice Media, 2019a).

In September 2022, Australia's parliament enacted legislation to raise the target for greenhouse gas emissions reduction to 43% below 2005 levels by 2030 (https://www.climatechangenews.com/2022/09/08/

australia-passes-climate-law-targeting-43-emissions-cuts-by-2030/). This law represents a 'symbolic' achievement, and HGA remains critical and skeptical of its implementation. An episode underscores this skepticism by posing the crucial question: "How will we reach that target?" The primary administrative policy intended to implement this law is the Safeguard Mechanism. Established on July 1, 2016, the Safeguard Mechanism mandates that Australia's highest greenhouse gas-emitting facilities maintain emissions below a specified limit or baseline. If a facility exceeds its baseline, it must take measures to manage the excess emissions (https://www.cleanenergyregulator.gov.au/NGER/The-Safeguard-Mechanism).

The first issue with the Safeguard Mechanism is its origin: it was designed and implemented by the previous Liberal government, which was widely criticized for its environmental policies. Climate action was a central promise of the Labor Party's campaign, leading many to anticipate a more radical policy shift under the new government. However, Labor disappointed many by essentially repurposing existing policies, which HGA critiques as "thankfully, the shit party set the bar so low we could come up with a shit-lite policy and you'd still applaud us for it" (see Juice Media, 2023b for an acerbic criticism of Labor's continuation of Liberal policies, that include approving new gas and coal projects, and how HGA directly addresses political players). This criticism underscores the perceived inadequacy of Labor's approach, suggesting that merely adjusting existing frameworks is insufficient to meet ambitious climate targets.

The second major flaw of the Safeguard Mechanism is its design, which appears to protect the coal and gas industry. Labor has reintroduced this policy, mockingly referred to as "S&M," suggesting it will "hurt the planet, make it all hot and wet, but you'll love it 'coz it's a Labor policy! Except it's not. Fuck you!" During this delivery, the presenter poses provocatively with a whip, bathed in red light, using imagery and wordplay associated with BDSM to emphasize the critique. Although "hot and wet" alludes to extreme weather events, the delivery's novelty and humor are aimed at capturing the attention of a progressive audience who might be swayed by Labor's strategy. The episode leverages this to caution Labor voters against being seduced by superficial changes, reminding them that a policy's label as a Labor initiative does not inherently ensure authentic environmental action. By foregrounding political persuasion as a key factor in environmental behavior, the ad warns that voters who prioritize environmental values might become complacent when their chosen party introduces ostensibly progressive legislation. However, political allegiance should not prevent critical examination of the policy's substance and effectiveness.

The episode continues its critical examination, with the presenter donning a lab coat and spectacles to underscore the scientific perspective according to which the most effective way to reduce emissions is to cease fossil fuel extraction. Nonetheless, the Safeguard Mechanism does not aim to restrict fossil fuel production; instead, it even facilitates new projects. It requires producers to "purchase and surrender Australian carbon credit units (ACCUs)" (https://www.cleanenergyregulator.gov.au/NGER/The-Safeguard-Mechanism) to offset a small fraction of their emissions. Given the $140 billion profit the fossil fuel industry earned in 2022, the cost of buying credits amounts to just 0.05–0.09% of their annual profit, or "approximately fuck all." The mechanism fails to limit the number of credits companies can purchase, allowing them to potentially increase emissions, offset them with credits, and still claim to be net-zero.

Carbon credits are the centerpiece of this policy, yet there are significant doubts about their effectiveness. The episode reveals that two of the four members appointed to an inquiry panel on the integrity of carbon credits have ties to the carbon credit industry. Despite this potential conflict of interest, the panel gathered and commissioned evidence, suggesting that carbon credits are "mostly bullshit," yet it concluded they are "essentially sound." Consequently, the policy fails to halt new fossil fuel projects, minimally reduces emissions, and relies on a questionable strategy to manage emissions. It effectively "distracts you with bullshit so our mates can keep making the ching-ching for the rest of this decade." The ad concludes with the biting remark, "new government, same climate shitfuckery" (Juice Media, 2023c).

HGA therefore seeks to make complex technical topics and obscure legal issues accessible and understandable for the purpose of encouraging audience engagement, particularly through participation in liberal democracy. For instance, the episode titled "Honest Government Ad|The Great Australian Bight" (Juice Media, 2017) aims to inform viewers about the environmental and economic implications of oil production in the Great Australian Bight and encourages political action to counter that. After reminding the audience of the destruction of the Great Barrier Reef, the episode highlights the Great Australian Bight as a pristine habitat for endangered species and a critical resource for the fishing industry.

However, the government is poised to allow "a bunch of eco-criminals" to drill for oil in this area.

The financial benefits of this project would primarily benefit corporations that pay minimal taxes, contribute little to job creation, exacerbate climate change, and pose major safety risks. The episode asserts that "even oil companies have forecast that if, I mean when, a spill occurs it'll pretty much endanger the entire Southern coast of Australia, along with all the fishing and tourism jobs." This message underscores that the project not only fails to provide tangible benefits to the public and local communities but also threatens their livelihoods. By emphasizing personal benefit, the episode appeals to audiences with a strong power motivation while also mitigating judgmental discounting by bridging the spatial and temporal distance that might lead to an underestimation of risks. After laying out the information and establishing the high stakes, the episode calls on its audience to tell the government to "GTFO."

*HGA* started with a strong belief in representative democracy in that it presumes that wrong policies could be corrected if a large enough number of people are informed that they are wrong and then demand action from their representatives or elect a different Government. It employs satire to expose the skewed dynamics of regulatory and political systems and address biases arising from political persuasion. It uses informational social influence and appeals to economic factors, racial and cultural identity, place and belonging as determinants of behavior to encourage people to protest against misguided policies through conventional mechanisms of representative democracy.

The same hope in the mechanisms of representative liberal democracy resurfaces in other episodes. The episode "Visit New South Wales!" opens with a humorous invitation to visit places with double entendre names like Dick Lake, Hooker Park, Glory Hole Cave, Rooty Hill, and Cumbooglecumbang. Early on, it advocates for preserving the last remaining coalition government among Australian states before upcoming state elections. The episode begins with a historical perspective, noting, "first stolen in 1788 [by the British Empire], this was once a small penal colony wracked by shitfuckery. Today, thanks to us, it's a thriving modern state wracked by shitfuckery." It provides ample evidence of alleged corruption before addressing environmental concerns. It claims that "by weakening land clearing laws to help dodgy developers, we've destroyed so much of [native forest and wildlife], we've made it into the top list of deforestation hotspots alongside the Congo and the Amazon." The episode criticizes policymakers for prioritizing financial interests of certain businesses over the welfare of different species, communities, and sustainable life, emphasizing that these policies have not boosted the economy: "our state forestry is losing millions, which means we're making people here pay to destroy their own native forests." This illustrates how citizens bear the costs of misguided decisions and policies. Besides, HGA endeavors to place contemporary environmental degradation within a historical context by highlighting how current policies and interests are influenced by legacies of colonial and imperial worldings. By drawing these connections, it underscores that the exploitation and environmental harm seen today are not isolated incidents but rather continuations of past practices rooted in colonialism and imperialism. This perspective emphasizes that the burden of these policies often falls on disadvantaged, disenfranchised, and dispossessed people, who pay the price for decisions driven by entrenched power structures and economic interests that prioritize profit over sustainable and equitable multispecies development.

The climate policy of the New South Wales Government is emblematic of its history of mismanagement and poor decision-making. The episode rebukes these policies by referencing a notorious moment when former Liberal Prime Minister Scott Morrison brought a lump of coal into the Australian Parliament to argue that coal is synonymous with economic prosperity and job security and falsely attributed opposition to coal mining and consumption to "ideological" and "pathological" reasons (2017, "Scott Morrison Brings in Lump of Coal Into House of Representatives" https://youtu.be/ea5bOaPkZpc). The episode features an image of this scene from 6 years earlier, with the HGA presenter sardonically commenting, "remember that time a lump of coal brought this piece of shit [i.e., Scott Morrison] into Parliament?" The ad then reveals that the New South Wales government mimicked this "stunt" in its parliament the following day.

As their climate policy failed to withstand scrutiny over time, the New South Wales Government attempted to rebrand itself as a climate leader. However, this rebranding effort unfolded as they approved 16 new coal mines during their last term in office and planned to sanction eight more coal mines by a company facing accusations of bribery and corruption. Despite these approvals, the most egregious aspect of their climate policy extends beyond coal investments. The ad provocatively states, "our biggest fuck-you goes out to the regional communities we claim to champion." This criticism is aimed at State Treasurer and Minister for Energy

Matt Kean, who granted Santos—an Australian domestic gas supplier and LNG supplier in the Asia Pacific region—approval to access farms without landowner permission. This decision facilitates the construction of an unwanted pipeline transporting coal seam gas from one of Australia's last great inland forests, directly beneath the fertile Liverpool Plains, by a company previously fined for contaminating aquifers with uranium. The episode punctuates this critique with a rhetorical question: "What could go wrong?"

The episode underscores the extent to which the government is influenced by the vested interests of corporations, while simultaneously failing to fulfill its responsibilities to protect citizens. Despite this critique, HGA maintains hope in the mechanisms of representative liberal democracy. Rather than merely listing grievances, the episode offers pragmatic solutions, urging voters to avoid re-electing the same government and advocating for the use of preferential voting, which remains optional in the state. HGA views preferential voting as a means to transcend the dominance of the two major parties, Liberal and Labor, and to introduce "not-shit" representatives capable of maintaining the last remaining coalition government in the country. The ad concludes with a call to action: "Visit New South Wales before we sell out what's left of it." This closing statement serves as a stark reminder of the urgency to protect the state's remaining resources and integrity (Juice Media, 2023d).[3]

In another episode addressing the 2022 floods, HGA reiterates its plea for its audience to vote, starting with a critique of Scott Morrison (Fig. 5.1). The episode highlights how the government ignored or suppressed numerous warnings about the severe risk of flooding. Despite the extensive damage caused by these floods, the government still plans to proceed with opening new gas fields, "which will ensure the next floods will be even worse." Rather than curbing the fossil fuel industry in light of the evident impacts of climate change, the government has granted support to the industry's continued destructive practices. This episode chastises the government's ineffective behavioral controls at the collective

[3] For further exploration of how oil and gas companies, like Santos, exert influence over Australian politics, see episode Honest Government Ad |Anti Protest Laws (SA) https://youtu.be/kecnSHmznic) and for advocacy on preferential voting, see episodes Honest Government Ad |Labor https://youtu.be/fHB0vDhdM3c; Honest Government Ad|Preferential Voting https://youtu.be/bleyX4oMCgM

**Fig. 5.1** Scott Morrison

level, emphasizing the disconnect between policy decisions and environmental urgency.

The government consistently supports the fossil fuel industry in the aftermath of disasters, deploying their "Big-Sack-of-Shit-in-Chief to the disaster area to take the heat for them [i.e., the fossil fuel industry]." This strategy is presented as part of the government's "Get-Fucking-Used-To-It Policy" for addressing climate disasters. Satire targets policies that aim to foster resilience against climate change and encourage communities to adapt to its devastating impact to serve the interests of the few. As Australia turns increasingly inhospitable due to environmental challenges, petroleum corporations continue to benefit from billions of dollars in subsidies and a tax-free status. Meanwhile, citizens who pay taxes and require rescue from their sinking homes are blamed for living in the path of what has been called "Chevron's trail of destruction."

While imposing a levy on fossil fuel exports could offer a partial solution, the government remains unwilling to pursue this option, preferring instead to make communities bear the costs of their losses: "We'll be counting the interest on the unspent disaster recovery fund, while you crowdfund choppers and rescue each other in tinnies." HGA argues that the solution lies in voting the Liberal Party out of office in the upcoming election "get a government that isn't captured by fossil fuel interests and who won't abandon and gaslight you as we head deeper into the climate crisis." In essence, HGA highlights systemic issues that prioritize the interests of a select few over the public good, effectively accusing the system of functioning as a kleptocracy, particularly in its assertion that the State is captured by fossil fuel interests. Despite these criticisms, HGA continues to advocate for active participation in representative liberal democracy as a means of altering the status quo. By encouraging voters to engage in the electoral process, the episode suggests that change is possible through existing democratic mechanisms, and that informed civic participation can challenge entrenched power structures and drive policy reform (Juice Media, 2022e).[4]

## Undefeated Despair over Kleptocracy

In the episode "Honest Government Ad|Carbon Capture & Storage," the theme of state capture or kleptocracy resurfaces, particularly in relation to the inefficacy of carbon credits, but this time democratic participation is not proposed as the solution. The government is again the primary target of satire, with the presenter, who is "from the Australian Government," announcing that we are entering the next stage of climate change, characterized by more frequent extreme weather events and disasters. "As things fall apart and calls grow for us to urgently reduce our emissions, we've come up with a PR campaign, I mean technology that makes us look like climate heroes whilst doing sweet nada." The satire targets inefficacious ecomodernist technofixes that not only fail to address the problem but also perpetuate the status quo. Carbon capture adn sotrage (CCS) is introduced with a logo reminiscent of Coca-Cola, emphasizing the superficial nature of such solutions and corporate complicity therein. The episode critiques these approaches as more focused on public relations than on implementing genuine solutions that require radical measures (Fig. 5.2).

Using visual aids, the presenter elaborates on the technology of CCS with capture and storage (CCS) with a satirical twist. CCS is described as "a complex mining process whereby fossil fuel companies inject donations into the arseholes of politicians to delay climate action and let them keep making the ching-ching." This definition begins with a seemingly serious

[4] For another episode that seeks to impact environmental policies through voting behavior see Honest Government Ad|Climate Emergency & School Strikes https://youtu.be/OL8a1YEhk_o

**Fig. 5.2** Coca-Cola

tone, only to pivot to bathos, enhancing the satirical impact. The visualization of this process amplifies the critique, providing a vivid commentary on the influence of fossil fuel interests.

The episode further delves into the stages of the CCS process. In the first stage, termed the "the capture phase, also known as State capture, fossil fuel interests infiltrate your government at every level, so that we'll ignore scientists and keep approving new coal and gas projects." As the presenter explains this phase, the visuals depict two politicians standing behind podiums in front of the seat of Government in Canberra. The politician with a blue tie, representing the Liberal Party, declares "we'll frack," followed by the politician with a red tie, representing the Labor Party, echoing the same sentiment. This imagery underscores the bipartisan complicity in supporting fossil fuel interests, highlighting the pervasive nature of state capture in delaying meaningful climate action.

The process continues with the "storage phase," in which the government pays fossil fuel companies billions of dollars to bury their emissions underground. Satire effectively illustrates the absurdity of the profit-driven cycle in a manner that is accessible to the audience. This is reinforced when the presenter rhetorically asks, "does it work?" and answers, "Absolutely!"—not referring to the efficacy or viability of carbon capture as a technological solution to climate change, but rather to the absurdity of a system that enriches politicians and the fossil fuel industry at the expense of people and the environment. The ad pointedly notes, "those companies bury the billions of dollars deep, deep down in their bank accounts where no tax officer can reach them."

Meanwhile, CCS has consistently proven inefficacious, having missed all its targets and milestones. To substantiate this claim, the ad presents data, but not in the conventional scientific manner. The presenter, now wearing a lab coat and glasses, utilizes effective graphics to make large numbers comprehensible to a non-expert audience. This approach ensures that the critique is not only clear but also resonates with viewers, highlighting how CCS serves more as a financial boon for fossil fuel companies than a viable solution for mitigating climate change.

The ad scrutinizes the performance of the Gorgon project in Australia, the largest CCS initiative globally, to highlight the inefficacy of carbon capture and storage technology. It notes that Chevron was given $60 million to capture its emissions, yet "in 2016 they captured fuckall. In 2017, fuckall. In 2018, fuckall. But in 2019 it finally got clogged up by sand." While Gorgon has managed to capture only a minimal amount of emissions, as little as its operating emissions, the ad sarcastically points out, "but hey, Chevron stored 100% of that $60 million you gave them."

This critique extends beyond Australia, illustrating a broader pattern of failure with CCS technology. Canada, for instance, has invested $1 billion in CCS, failing to meet its targets every year. Similarly, the US government falsely claimed that CCS could lead to "clean coal." However, after years of delays and $7.5 billion in expenditures, the plant suspended operations, and another plant in the country met the same fate. These examples collectively emphasize that CCS technology has not only been ineffective but has also served as a costly distraction from more viable solutions to climate change.

CCS has not proven to be a viable solution, not only due to technological limitations but also because of its prohibitive costs and questionable benefits. The episode highlights that "CCS is so expensive that even if it worked, which it doesn't, implementing it would be unprofitable for our mates. Unless of course we make you pay for it, which is exactly what we've been doing by blowing $4 billion of your tax dollars on CCS." Despite the consistent failure of CCS projects, the Australian government continues to approve new ones. Moreover, these projects are funded by diverting resources away from agencies intended to support genuine climate solutions, such as renewable energy and electric vehicles.

This diversion of funds underscores a significant misallocation of resources, where the government invests in an impractical technology instead of viable alternatives. By channeling tax dollars to fossil fuel companies for emission reduction efforts—companies that "pay fuckall taxes in the first place"—the government effectively prioritizes non-viable and inefficient methods over sustainable, long-term solutions. This situation illustrates a systemic issue in which public funds are misused to uphold the interests of the fossil fuel industry at the expense of communities and meaningful climate action.

The situation is indeed absurd and increasingly desperate, as political systems, economic interests, and technology collectively shape environmental behavior at institutional and organizational levels. Despite its clear irrationality, the government and fossil fuel companies manage to market CCS through enhanced oil recovery (EOR). This process uses captured $CO_2$ to extract oil from hard-to-reach deposits, ultimately leading to more carbon emissions than those initially captured. The ad wryly notes, "you might call that bullshit, but we call that carbon capture use." This commentary highlights how the government and fossil fuel industry manipulate framing and public discourse to mislead the public for economic and political gain.

This manipulation is particularly egregious because it diverts time and resources toward a technology that fails to mitigate climate crises, thereby hindering progress by redirecting resources away from effective and available solutions. The same strategy is employed to obscure the negative impact of hydrogen extracted from fossil fuels. While hydrogen produced through electrolysis using renewable energy is a zero-emission energy source, the ad points out the irony: "Here's clean Hydrogen, made from coal or gas which emits more than if you just burned the fucking coal or gas in the first place." The government continues to label this as 'clean' while burying the problem using the same ineffective CCS technology (Fig. 5.3). This approach underscores a systemic issue where misleading narratives are used to perpetuate environmentally harmful practices under the guise of innovation, further delaying necessary and meaningful climate action.

The episode concludes with a sharp critique, stating, "and those are just some of the New Technologies we'll be using to avoid taking real climate action. And when you add them to the mix [of CCUS], you can clearly see what we're all about [i.e., CCUNTS]." This closing remark underscores

**Fig. 5.3** Carbon capture and storage

the satirical portrayal of the government's reliance on ineffective technologies as a means to sidestep genuine climate commitments. The closing credits read "Australian Government: CCUNTS," further emphasizing the episode's message that the government's strategies are more about maintaining appearances while serving corporate interests than enacting substantive environmental reforms (Juice Media, 2019a).

Another episode tackles the desperate situation by illustrating how the government manipulates resilience and adaptation to cover up its failure to take timely and proportionate action against climate change. While different governments have largely failed to implement effective climate strategies, adaptation and resilience have recently emerged as their preferred alternatives to genuine solutions. In the aftermath of the catastrophic 2020 wildfires in Australia, HGA explores the efficacy of these approaches. The episode opens with a welcome to the Anthropocene, an era characterized by fire, floods, blame shifting, and misrepresentation. In response to the disastrous fires, the Australian Government has "decided it's time to take action. Introducing our new and updated climate policy: Get Fucking Used to It!" Rather than mitigating the crisis, the government uses adaptation and resilience to imply that suffering and death due to climate change are inevitable, urging people to accept this new reality.

The episode develops the metaphor of an abusive relationship by comparing this policy to gaslighting a nation by maintaining "it's all part of the new and exciting stage of our abusive relationship with you, where rather than ceasing our shitfuckery, we tell you your only choice is to adapt to it." While climate policy is reduced to adaptation and resilience, the government invests in boosting gas consumption. Consequently, the ad concludes, "as things fall apart—and let's be clear this summer was just a taster—you might want to start organizing your own communities, 'cuz you're on your own, dickheads. We'll be in Hawaii sending thoughts and prayers." This statement underscores the notion that the government prioritizes the safety of the fossil fuel industry over its citizens. It also underlines the efficacy of organizing at the community level as opposed to praying.

Despite the damage caused, the government continues approving new coal and gas projects to ensure industry profits. It manipulates and misrepresents statistics to justify supporting the fossil fuel industry, all while claiming Australia is "the best-placed nation to lead on climate action." However, the government is "led by a man steeped in a delusional apocalyptic cult," hindering the country's potential for climate leadership. Scott Morrison, a fervent Pentecostal, has infamously claimed he was called to do God's work (Martin, 2021). "The mission of Morrison's Pentecostal faith—a cult that espouses the Seven Mountains Mandate—is literally world domination. This is the unequivocal and openly stated primary Pentecostal aim" (Pini, 2022). Besides place and belonging, as well as political system, therefore, the episode underscores how belief in divine solutions can hinder pro-environmental action by highlighting religious identity as a determinant of environmental behavior. The cult's Seven Mountains Mandates—i.e., education; religion; family; business; government and the military; arts and entertainment; and the media—clearly prioritize material success and imply cornucopia (Garrard, 2023) over sustainability and establishing symbiotic relationships (Haraway, 2015, 2016). Hardly does this inspire hope or participation in neoliberal democracy as a way out (Juice Media, 2020d).

## Kleptocracy, Neoliberalism, and Capitalism

HGA is fully aware that it is neoliberalism and capitalism that lead governments to fall into the precipice of kleptocracy. In the episode "Honest Government Ad|Net Zero (feat. Greta Thunberg)" (Juice Media, 2021b),

HGA explores the intersections of neoliberalism, capitalism, and kleptocracy. The episode begins with satire targeting the government by placing a trademark symbol over the word "Government™." This symbol, typically associated with corporations and profit rather than public institutions committed to the collective good, suggests that the government operates more like a corporation aiming to secure profit for a small group of investors at the expense of its citizens. Besides, the government logo features a camera atop an iconic governmental building, which can be seen as a critique of the compatibility between liberal democracy and mass surveillance.

Foucault's (1977, 1991) concept of governmentality provides insight into this critique. Foucault argued that modern states employ surveillance and other disciplinary mechanisms to regulate populations, ensuring conformity and control. Surveillance is integral to the functioning of neoliberalism and capitalism, that emphasize market freedom and individual responsibility. However, it may seem paradoxical that a representative democracy, which values participation and freedom, including free market, resorts to mass surveillance to maintain its existence. A government truly by the people and for the people should not need to position itself against its citizens, highlighting the tensions between democratic ideals and the realities of neoliberal governance.

The episode draws attention to the misplaced trust people have in governments to address the climate crisis "so humanity can keep enjoying its favorite pastime: continuing to live on this planet." It seeks to convey the urgency of the situation and counteract judgmental discounting, where risks are underestimated due to their perceived temporal distance. However, the episode argues that the government is part of the problem rather than the solution. To illustrate the severity of the crisis, the presenter uses a "graph shaped like a penis 'cuz it shows how fucked we are," explaining the critical stakes of surpassing a global warming threshold of 1.5°C above pre-industrial levels, or "the-stop-here-or-we're-fucked point."

The ad employs straightforward language and visual aids to demystify complex issues, informing its audience that the current trajectory will not achieve the 1.5°C target. The episode critiques governmental promises, often thwarted by corporate power or reduced to empty words in "planphlets," and warns of reliance on non-existent or inefficient technologies like carbon capture. It argues that the only genuine solution is to cut and eventually cease carbon emissions. Yet, this solution lacks support from a government that subsidizes the fossil fuel industry at $11 million per minute, distorting the market and deterring renewable investments.

The government, the ad suggests, cannot afford honesty because it would necessitate implementing viable solutions like ending subsidies and banning fossil fuel projects. Instead, it opts for the ambiguous "Net Zero by 2050" pledge. The presenter explains the stakes of this plan. Despite the promise of net zero, the trajectories differ significantly, potentially resulting in three times more emissions before 2050 (Fig. 5.4). Ultimately, the government reduces its response to an "exciting experiment with the only planet that we know supports life in the whole fucking universe just so we can make a few billionaires even richer." This critique underscores the dangers of prioritizing short-term economic gains over long-term sustainability.

The episode brings its critique full circle by returning to the notion of the government as a corporate entity, stating it is "captured by the fossil-fuel industry, compromised by … [its] moral failings and lack of vision, and probably going to get us all killed." This framing underscores the government's entanglement with corporate interests, suggesting that such entanglements hinder genuine climate action. The "local Government™ franchise" ad concludes with a cameo by Greta Thunberg, who appears on screen to deliver the final satirical jab: "authorized by the Department for

**Fig. 5.4** Net zero emission

Blah Blah Blah.' This closing remark highlights the perceived emptiness and performative nature of governmental commitments, reinforcing the episode's critique of how superficial rhetoric is prioritized over substantive environmental policy.

The episode "Honest Government Ad|Renewable Energy" critiques the government as the primary obstacle to transitioning to renewable energy. It argues that the government has avoided implementing policies that would facilitate this transition and instead promotes the National Energy Guarantee (NEG), described as "such a watered-down pissweek policy that it's literally worse than doing nothing at all which is precisely why our BFFs [i.e., "Top fossil fuel donors: Minerals Council of Australia, Chevron, Woodside, Origin, Santos, Wesfarmers"] love it so much."

The ad further challenges the neoliberal argument that government intervention should be minimized, allowing the market to dictate energy preferences. Contrary to this laissez-faire approach, the government is accused of actively supporting a declining industry using public funds. It is actively "propping up a dying industry with your tax dollars or trying to force this ancient relic [i.e., Liddell Power Station, coal-fired thermal power station] to stay open, when even its owners are saying coal is dead." Through this critique, the ad exposes the government's betrayal of its own claims by manipulating the market and intervening in ways that contradict the principles of free market capitalism. It highlights how the discourse of laissez faire and free market capitalism and the latent contradictions in governmental behavioral controls shape environmental behavior at institutional and organizational levels.[5]

While the Federal Government lags in leading the transition to renewable energies, some states have demonstrated leadership and strategic foresight by investing in renewables, and are now reaping the benefits. In contrast, the Federal Government is accused of investing in "Bullshit," described as "what fuels our daily fearmongering about blackouts and high prices when in fact renewables are responding to peak demand and reducing electricity bills." This satire highlights the government's reliance on misleading narratives, despite contrary evidence that renewables are reliable and affordable. By challenging these false claims, the ad seeks to

[5] For a similar satire on the hypocrisy of free market capitalism, "class warfare," and "corporate welfare" in the context of environmental justice, see "Honest Government Ad|Adani" https://youtu.be/G1SdCNdPDL8

dismantle the government's fear-based rhetoric and emphasize the tangible benefits of renewable energy adoption.

The episode goes further by aiming to establish and reinforce norms, highlighting that the market has broadly embraced and invested in renewables. After enumerating several market segments that have made significant investments in renewable energy, the presenter notes, "even these pricks [i.e., Adani] are investing in solar." She questions rhetorically, "and why not? With the prospect of new jobs, cheaper bills, and avoiding epic climate change fail, what kind of energy Neanderthal would want to stand in the way of the renewables boom that's coming?" The rhetorical question is answered with "us, the Australian Government," accompanied by an image of Scott Morrison holding a lump of coal in parliament. The episode highlights personal financial interests, emphasizing the benefits of affordable and secure energy, improved economic prospects, and increased employment opportunities to appeal to those motivated by power and economic gain. By doing so, it seeks to engage stakeholders who prioritize financial incentives.

The episode concludes with a satirical twist: "The Australian Government: We've stopped the Boats; It's time to Stop the Sun." This closing remark references the government's past use of fear-mongering about illegal immigration to gain political favor, drawing a parallel to its current strategy of invoking fear around renewable energy. The absurdity of "stopping the sun" underscores both the inevitability of the renewable energy transition and the folly of government plans that oppose or hinder its progress. This critique serves to highlight the irrationality of resisting renewable energy advancements and the need for a shift in governmental policies.

Detailing the environmental and economic benefits of renewable energy, the ad applauds the new era of opportunity and criticizes politicians who fail to capitalize on it. It underscores the dual advantages of renewable energy—environmental sustainability and economic growth—and calls out the government for hindering progress due to outdated allegiances and policies. Through this critique, the episode seeks to encourage a shift in both policy and public perception towards embracing renewable energy as an inevitable and beneficial transition. In doing so, however, it falls back on market capitalism and neoliberal discourse (Juice Media, 2018).

HGA also targets corporations, often those responsible for environmental degradation. In the episode "Honest Government Ad|Visit the Northern Territory!" HGA critiques fracking practices and targets corporations involved in the process. The ad provocatively introduces gas—not the iconic Uluru—as the "greatest natural treasure" of the Northern Territory. Using satire to inform viewers about the fracking process, the ad claims, "fracking is a complex mining process whereby gas companies inject donations into the arseholes of politicians, to release access to land so they can tap into shale and coal seam gas deposits." By citing expert and scientific studies, the episode underscores the dangers of fracking, asserting that it contaminates water, pollutes air, and endangers children's health. The visuals of tap water and a river on fire concretely illustrate these concerns. The episode further emphasizes the precariousness of the situation by noting that "the NT [i.e., Northern Territory] is littered with decades of clusterfucks in environmental mismanagement," listing issues such as leaks of radioactive materials, acid, copper, and PFAS toxins.

The episode critiques the safety and economic rationale of the fracking project in the Northern Territory, highlighting that it creates only 524 jobs, described as part of the "booming revolving door industry." This is visually represented by an image of a revolving door labeled with "government jobs" in green and "mining/energy jobs" in red. The ad underscores the environmental impact by stating that "emissions from NT's McArthur basin would dwarf those from Adani coalmine" (Davidson, 2017), being the "equivalent of building 130 coal power plants and operating them for 40 year," thereby making the scale of emissions more tangible for viewers.

Despite the project's significant breach of the Paris Agreement, the government continues to advance it under the guise of addressing a non-existent gas crisis. While Australia is abundant in gas deposits, the industry opts to export gas due to higher profitability, depriving Australian citizens of the economic benefits. The ad asserts, "'cuz you know, the real gas crisis is that these cunts [i.e., Chevron, Statoil, Santos, Origin, Hancock Prospecting, Halliburton] run your Government." Instead of imposing caps on gas exports and company profits, the government pressures the Northern Territory into fracking, and "fucking public health [we see the video of a little girl drinking water], farming and indigenous communities, the world-famous tourism industry, and the climate all in one go" (https://youtu.be/cuQZ6rALub8?si=6bJQMYGSA7O-LKVm). The episode therefore demonstrates the concept of differentiated responsibility, highlighting how governments, under a kleptocratic system, often prioritize serving corporations over their citizens. This situation illustrates how economic systems are structured to benefit the wealthy disproportionately and usually at the cost of the less privileged communities, emphasizing the alignment of governmental actions with corporate interests at the expense of public welfare and environmental sustainability.

Another episode focuses on Woodside Energy Group Ltd, an Australian petroleum company, with the presenter sarcastically introducing herself as, "hello! I'm from Woodside. I mean the Western Australian Government." This introduction implies a significant conflict of interest, suggesting that the petrochemical corporation holds substantial influence over governmental operations. The episode begins with a seemingly respectful "Acknowledgement of Country," intended to honor the traditional custodians of the land. However, the presenter quickly undermines this gesture by pointing out Woodside's "shitty logo [appears] everywhere: on the city, footy [i.e., football] team, festival, uni', theater, orchestra, ballet, scitech, even on children." Visual evidence for each claim is provided in the background.

The episode satirically notes, "kids will need to swim a lot to survive the worsening floods we're causing, so Woodside's helping them to learn young," highlighting the irony of the company's support for swimming lessons amidst sea level rise and more frequent and intense flooding it contributes to. HGA questions the motivations behind Woodside's financial support, suggesting that the primary drive is vested interest rather than genuine community benefit. The company gains free publicity and a positive public image, while the ad cynically adds, "plus we're sadists, so using your child as a billboard of the company that's trying to kill them gives us wood," accompanied by an image of a child wearing a Woodside-branded bathing suit. This commentary critiques how corporate sponsorships and public relations efforts often mask underlying self-serving agendas, challenging viewers to reconsider the true intentions behind corporate philanthropy. The use of children as a framing strategy in the ad is particularly effective, as it resonates with a broader audience, including conservatives who may be indifferent to environmental issues or skeptical of climate change. By focusing on the corporation's impact on children, the message shifts from abstract environmental concerns to tangible effects on what is personally significant and cherished by people across the political spectrum. This approach leverages emotional appeal, making the issue more relatable and urgent, and encouraging viewers to reconsider their stance on corporate practices and their broader implications.

After targeting Woodside, the episode shifts its criticism to the state government. Through simple graphics and satire, it informs viewers that Western Australia is the only state in the country witnessing rising emissions, primarily due to Woodside's repeated disregard for the Environmental Protection Agency (EPA)'s recommendations to reduce emissions. The Western Australian Government's support not only has enabled Woodside to ignore the EPA but has also led to the approval of the Burrup Hub, described as "Australia's most polluting project ever." It also enables Woodside to open two new mega gas fields, "expand this lawn sausage [i.e., Woodside PLUTO LNG Terminal] and extend this tumor's life [i.e., North West Shelf Plant] by half a century." This will emit twelve times the country's annual emissions, severely undermining global efforts to reduce emissions in line with the Paris Agreement. The critique reaches its peak with the satirical remark that this trajectory will "drive us closer to what local philosopher OzzyMan calls Destination Fucked." By focusing on both the corporation's and the government's roles, the ad underscores systemic failures in environmental governance and the urgent need for accountability and reform.

The significant and apparent negative environmental impacts of the project prompt the question of why the government proceeds with it. According to HGA, the answer lies in the benefits reaped by big oil companies, including Chevron, BP, Shell, BHP, and Woodside, which will profit from this lucrative project while "paying bugger-all royalties." As this is conveyed, the chyron reads "Rubbing Australia," underscoring the notion of exploitation. The ad highlights that the government raises less revenue from gas royalties than from vehicle registration fees. Furthermore, "most of these companies also pay bugger-all taxes, are mostly foreign owned, and employ fewer people than any other sector," with each assertion supported by evidence displayed in the background.

HGA is clearly aware that the use of humor and satire can undermine its credibility, so it addresses this concern by providing references and visual evidence to bolster its claims. This approach not only enhances credibility but also ensures that the serious message is effectively communicated alongside the related satire. By juxtaposing satire with factual support, HGA aims to engage viewers while maintaining the integrity of its critique.

The economic benefits of the project for Western Australia and its communities are minimal. The government cites energy security as a justification for proceeding with the project, despite its numerous drawbacks.

HGA counters this rationale by stating, "well, renewables would do that too, minus the floods and fires, but here we are delaying them by investing in shit like this." This response exposes another layer of the government's indecency and hypocrisy.

Moreover, most of the gas produced at the Burrup Hub is intended for export, ostensibly to "help Asia adopt renewables." This claim is based on a report commissioned by Woodside and completed by The Commonwealth Scientific and Industrial Research Organisation (CSIRO). However, Woodside later refused to release the report when its findings did not support the company's unfounded claims.[6] By pointing out these inconsistencies and the lack of transparency, HGA underscores the government's misplaced priorities and the detrimental effects of prioritizing short-term economic gains for a few over sustainable solutions.

The project's drawbacks extend beyond environmental damage and minimal economic gain to include significant safety concerns. It involves "drilling under the Scott Reef, displacing whales with seismic testing, and risks rooted everywhere from Ningaloo to Timor if there's a spill." HGA addresses both prevention and promotion goals by highlighting not only environmental and safety risks but also touching motivational types in exposing the project's negative cultural consequences. It warns of the potential destruction of the nearby Murujuga Rock Art, one of the oldest art galleries in the world, by referencing a past incident: "Remember that time our mates blew up this ancient sacred site?" This alludes to the Rio Tinto Group's destruction of an ancient Aboriginal sacred site, invoking memory as a powerful determinant of behavior. This dual appeal to both preventing further loss and promoting the preservation of invaluable cultural heritage strengthens the ad's call for reconsidering the project's approval.

After presenting compelling reasons against the project on various levels, the presenter poses the question, "Are we gonna do it anyway? You betcha! We are just waiting for approvals from Canberra." In encouraging the audience to contact their federal MP, the episode focuses on mobilizing public action to influence policymakers rather than changing individual behavior. The ad targets systemic issues, listing connections between

[6] On the conflicting interests of CSIRO, Bureau of Meteorology, and the fossil fuel industry in the form of Gas Industry Social and Environmental Research Alliance (GISERA) see "Honest Government Ad|We Make Everything Good Sh!t" https://youtu.be/-uXo7wtGW7M

governmental figures and Woodside: "just 'cuz our former treasurer went straight onto the Woodside board, as did our former Federal Resources Minister, and just 'cuz the last three chairs of our regulator used to work for Woodside, and our premier threatened to bypass the courts if they rule against Woodside, and just 'cuz four of our last five Premiers had gas lobby staff working in their offices, and gas companies donate shittons to both major parties [i.e., Labor and Liberal], Woodside being the biggest donor of all." While the voiceover narrator lists these, the presenter interjects, finally finds a chance to chime in, "doesn't mean that the gas industry has captured this government. It just means it is the Government.... Whatever, we own this place. Fuck you." This blunt conclusion underscores the ad's critique of kleptocracy, suggesting that the gas industry effectively runs the government. The ad claims to be "Authorized by the Department of State Capture," reinforcing its message of systemic corruption and the urgent need for public intervention (Juice Media, 2022e).[7]

While the HGA expresses frustration at the pervasive influence of the petrochemical industry on the government, it still encourages audiences to contact their MPs. Although the tone conveys disappointment, the series stops short of advocating for radical change or alternative systems. Instead, it operates within the framework of representative democracy, urging public engagement to instigate policy change. However, a more radical stance is found in another episode that advocates for decarbonizing the economy. This episode calls for a fundamental shift in economic priorities, emphasizing the urgent need to transition away from fossil fuels and invest in sustainable energy solutions. By promoting decarbonization, the ad challenges viewers to consider more transformative approaches to addressing climate change, reflecting a bolder vision for achieving sustainability.

Like many observers, HGA viewed the recovery from the pandemic as a pivotal opportunity to initiate a sustainable and equitable transition. However, the Australian Government opted to facilitate economic recovery through support for the gas industry. The Disaster Economic Recovery Plan (DERP) is satirically described as being "brought to you by the empty ballsack of ideas that is Scott's [Morrison] Ministry." The government frames investments in the gas industry as a means to facilitate a transition to a sustainable future. In contrast, HGA counters that gas "fuels

[7]For further exploration of Woodside's political influence, see "Honest Government Ad|Visit Timor-Leste!; (https://youtu.be/xqegTsi6SiE)

the transition of dollars from public coffers into the bank accounts of private companies who pay no tax and create shittons of emissions."

The episode makes clear that reliance on gas is not endorsed by independent experts, does not generate new jobs, nor does it lower energy costs for Australians. While experts assert that renewables outperform fossil fuels, Angus Taylor, referred to as "The Minister for Energy & Emissions Reduction (LOL)," dismisses evidence, proposing instead to "fuck evidence! Let's invest your money in bullshit solutions like carbon capture which no credible study says will ever work, but which gives his mates [the logos of Origin, ExxonMobil, Chevron, Santos, Woodside, Rio Tino, Adani, and Minerals Council of Australia appear on the background] a boner 'cuz it let's them keep polluting and making the ching-ching." This critique underscores the disconnect between governmental narratives and expert recommendations which seems to serve corporate interests.

HGA critiques not only the environmental costs of the government's recovery strategy but also its implications for social justice. It draws parallels to President Franklin D. Roosevelt's New Deal, which addressed the Great Depression by employing people in infrastructure projects and expanding the social safety net. HGA advocates for a similar approach in Australia, suggesting that decarbonizing the economy could both invigorate the high-tech job market and benefit the planet, promoting a holistic approach that intertwines economic and planetary well-being.

Instead of embracing a comprehensive vision, the Australian Government has chosen to "subsidize an industry that manufactures fuck-all apart from holes in the ground and continent-scale bushfires," neglecting support for artists, casual workers, universities, and job seekers. HGA emphasizes that sustainability and socio-economic justice should not be seen as separate initiatives. The episode concludes by criticizing the government's choice to "renovate rich people's homes instead of public housing and those who lost theirs in fires," highlighting the misalignment of priorities in addressing both environmental and social challenges (Juice Media, 2020b).

The furthest HGA takes its critique of the entanglement of representative liberal democracy with kleptocracy and capitalism is to encourage grassroots movements as a form of resistance. "Honest Government Ad|We're Fine" delivers a message from the Government™, sarcastically claiming "we are fine" despite widespread environmental disasters, such as fires in the Amazon, half of Africa, the Arctic, Indonesia, Spain, Greece, and even Greenland. The presenter says "we're fucked" to describe the current stage of climate change, emphasizing that these impacts will unfold "in your lifetime." This use of accessible language and slang serves to underscore the severity of the situation and to counteract judgmental discounting, which occurs when individuals perceive threats as distant and thus less urgent.

The ad attributes the current climate crisis in part to "wasting decades pissfarting around at climate summits with non-binding emission targets, whilst handing out subsidies to climate criminals, obstructing renewables, and generally not giving a shit that rising $CO_2$ levels are about to trigger what scientists call 'Feedback Loops.'" It explains this term in simple language, using visual aids to make the concept accessible: "the scientific term for when a species uses its own ignorance to screw itself and everything else around it so hard that its own planet tells it to GTFO" (Fig. 5.5).

The accompanying image depicts Homer Simpson, a character known for his lack of intelligence in the iconic show *The Simpsons*, holding a lump of coal, a deracinated tree, money, and a remote control. He is humorously portrayed as the Vitruvian Man, a drawing by Leonardo da Vinci that illustrates the ideal human proportions. Homer Simpsosn, however, is the last name that comes to mind when thinking about golden proportions. The juxtaposition of Homer with this iconic image highlights the

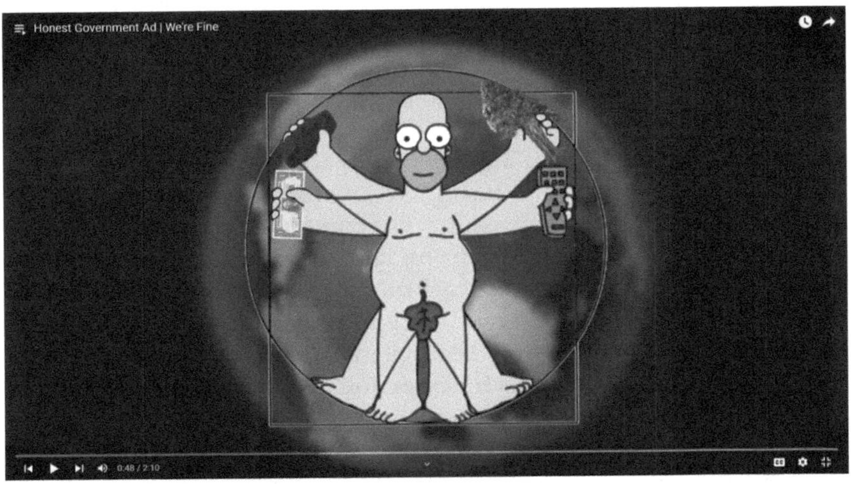


**Fig. 5.5** Homer Simpson as the Vitruvian Man

absurdity of human overconsumption and environmental degradation. The objects he holds in his hands symbolizes how Homo sapiens have extracted and depleted natural resources for manufacturing consumer products and economic value. The shape of the figure alongside the idea of overconsumption might remind the wary viewer of the doughnut economy, an economic model that advocates for sustainable development within planetary and social boundaries, emphasizing the need to balance essential resources for all people within ecological limits. Ultimately, the ad's message is clear: humanity has failed to regulate its impact on the planet, and immediate action is required to address this imbalance.

The episode highlights the dire consequences that many Pacific nations are already experiencing due to rising sea levels, emphasizing Australia's indifference to their pleas for reducing emissions. This plight serves as a stark reminder of the tangible impacts of climate change on vulnerable regions unfolding in real time. The advertisement cautions that the current trajectory may lead to armed conflicts over access to essential resources such as water and food. It adds a satirical jab at the United States, suggesting that "except for [US] America where the Chosen One will just nuke hunger," directly criticizing the US as the largest emitter of $CO_2$ and condemning its inadequate policies and leadership in addressing an existential crisis.

The ad further underscores how governments and corporations manipulate emotions to deflect blame from themselves onto individuals. It sarcastically suggests, "please don't panic. If the realization that we're fucked troubles you, why not ride a bike to work, have shorter showers, or sent thoughts and prayers? Just don't join the global climate strike … or Extinction Rebellion … 'cuz a sustained, collective movement would force us to take drastic action and turn this ship around, which might just be doable if enough of you demand it." The episode therefore acknowledges the impact of eco-anxiety while exposing how policymakers and corporations exploit it. They propose trivial solutions that have minimal impact relative to the scale of the existential urgency, thereby shifting responsibility to citizens. Furthermore, the suggestion of sending thoughts and prayers illustrates the gap between intention and action, emphasizing that belief in divine solutions can act as a psychological barrier to pro-environmental behavior (APA, 2009; Zekavat & Scheel, 2023). The ad suggests that the antidote to this manipulation is for people to demand immediate and proportionate action from governments to genuinely address the environmental crisis. This requires collective mobilization and advocacy for policy changes that match the magnitude of the crisis.

While the episode maintains a thread of hope, its tone is not overtly optimistic and concludes on a somber note: "Authorized by the Department of Going Gentle Into That Good Night." This is an allusion to Dylan Thomas's poem "Do not go gentle into that good night," in which the speaker implores a silent interlocutor—often interpreted as the poet's father—not to succumb to death without a fight. In contrast, HGA uses the affirmative form, suggesting that the government is effectively leading everyone toward destruction. The ad implies that without significant civil resistance—which "might just be doable"—the trajectory towards environmental and societal collapse will continue unchecked. This closing statement reinforces the urgent need for collective action to counteract governmental inertia and redirect efforts toward sustainable solutions (Juice Media, 2019b).

## Environmental Justice

Besides advocating for grassroots activism, HGA occasionally addresses environmental justice issues. The episode "Honest Government Ad|Adani Coal Mine" (Juice Media, 2016) critiques the Australian Government's priorities in investing in coal and railways rather than in renewables and broadband. The advancement of the "Carmichael Coal And Rail Project (CCRAP)," the largest new coal mine in the world, aligns with policies that stoke fear against renewable energy. These policies propagate false claims, such as suggesting that renewables cause lightning and child drowning.

The episode critiques not only the regressive nature of the technology involved but also the colonial implications of the project. By "disregarding the protests of the Wangan and Jagalingou traditional owners who, just like those pesky natives at Standing Rock, are standing firmly in the way of our CCRAP to defend their sacred country and waters from obliteration," the government perpetuates colonial practices, highlighting the intersection of environmental and racial justice. The project's negative environmental impacts extend to significant landmarks like the Great Barrier Reef and the Great Artesian Basin, contributing to climate change by emitting 4.7 billion tons of $CO_2$, which constitutes 0.5% of the remaining global carbon budget to limit warming to 2°C above pre-industrial levels.

Furthermore, the Australian government has contracted the project to Adani, a company with a history of human rights violations, environmental destruction, and corruption. Satire in this episode serves to inform the audience, underlining the interconnections between various forms of oppression, and criticize governmental policies. The call to action appears in the epilogue, urging viewers to "stand with the Wangan and Jagalingou to Stop CCRAP," reinforcing the need for collective resistance to protect both environmental and cultural heritage.

In another episode, HGA offers a nuanced discussion of carbon credits and offsets—strategies favored by governments and businesses to support their net-zero claims to critique Carbon Colonialism. The episode begins by defining carbon credits, using a practical example to ensure the audience's understanding. Instead of a typical academic definition, the ad shows a webpage from an "Airline," modeled after Qantas' website without the brand name, explaining how carbon credits are used to offset flight emissions. This common scenario serves as a relatable entry point into the critique.

The ad swiftly launches into criticism, inviting viewers to "imagine that your money was actually being spent on strippers and coke, except the strippers are our mates"—with logos of companies like Origin, Chevron, Woodside, and Santos appearing—and the "coke is gas." This metaphor suggests that resources allocated to offsets are squandered and highlights that these resources come from "*your* money." By framing the issue in terms of vested interests, the ad raises the stakes and personalizes the message, bringing financial risks closer to home.

The realization that voluntary contributions intended to reduce one's carbon footprint might actually exacerbate the problem fosters a sense of betrayal. Betrayal, as an uncertainty emotion, can enhance systematic reasoning and elaboration, potentially leading to stronger attitudes (Strick, 2021). By portraying offsets as a personal betrayal of trust, the episode aims to engage the audience more deeply, encouraging them to think critically about the issue and amplifying the ad's impact.

The episode continues to delve into the workings of Australian Carbon Credit Units (ACCUs). Initially, it presents the government's narrative, which paints a rosy picture: "there's just one tiny problem: we may have been handing out millions of ACCUs for bullshit like paying landowners to grow trees that are already there, not clearing land they were never going to clear, and growing forests where forests can never grow." As a result, 70% to 80% of ACCUs lack integrity, meaning they do not provide additionality. The fossil fuel industry embraces these "fuckoffsets" because they are inexpensive, allowing companies to afford them and 'offset' new projects.

The ad illustrates how carbon offsets are failing to fulfill their intended purpose of offsetting unavoidable emissions due to a lack of "transparency, integrity, and rigorous accounting." It alleges that the government has compromised all three by appointing individuals with vested interests to oversee offset programs. Angus Taylor, the Minister for Energy and Emissions Reduction, previously involved in document forgery and other controversies, was "in charge of this gig." The Emission Assurance Reduction Committee was directed by an expert until he "started asking too many questions, so Angus replaced him with this fossil fuel lobbyist."

Moreover, the ad claims that the integrity of offsets is further weakened by employing a person "whose dodgy modeling we used to bring you bullshit fear campaigns" and the former CEO of Origin Energy to review carbon credits. Fossil fuel companies like Santos can earn ACCUs through carbon capture, a technology notorious for missing its targets. The ad argues that by undermining these essential features of offsets, their effectiveness is severely compromised, rendering them ineffective in achieving genuine emission reductions.

The regulation of ACCUs is so lax that the Australian Government is anticipated to soon exhaust its capacity, prompting it to "export this shitshow to the whole Indo-Pacific with help from Woodside, you know, the pricks that robbed Timor-Leste of its resources." This expands the conversation to include themes of environmental justice and colonialism. The ad notes that "Fiji and PNG have already signed up for the next kick in the face: Carbon Colonialism, where we pay poor nations who've begged us to stop digging up fossil fuels to help us dig up even more fossil fuels."

By contextualizing carbon offsets within broader solutions that pursue endless growth and by linking these to colonial practices, the episode strives to adopt a holistic critique. It subtly, yet resonantly, critiques the complicity of formerly colonized nations in enabling the colonizers, despite the existential threats such actions pose to themselves. This critique of complicity highlights the complex dynamics at play in global environmental and economic justice issues.[8]

[8] For the intertwinedness of economic injustice, environmental justice, and the conflict of interest of fossil fuel corporations, governments, people, and future generations, see "Honest Government Ad|Dakota Access Pipeline" (https://youtu.be/a9TR9G5bd7w)

The episode succinctly encapsulates its critique by reminding the audience of their vested interests, stating, "you pay for the carbon credits, we make them as shit as possible so our mates can buy them for peanuts to 'offset' their emissions, so more gas basins get drilled, emissions keep rising, our neighbors get screwed again, and your town gets flooded again." This summary underscores the systemic flaws and the cycle of exploitation perpetuated by inadequate carbon credit regulations. The ad cautions viewers that even a Labor Government is unlikely to resolve these issues, as they too rely on offsets for their climate policy. It emphasizes that the only genuine solution is to conduct a "full inquiry into this shit-buffet" (Juice Media, 2022a).

## Conclusion

Not as deeply enmeshed in the constraints of traditional forms of the culture industry, such as TV and cinema, *Honest Government Ads* operates with a degree of freedom on the decentralized World Wide Web. Although it maintains a certain level of financial independence, its operations still rely on marketing merchandise, patronage, and ad revenues from platforms like YouTube. Arguably, this relative freedom allows the series to engage with a wider range of negative and uncertain emotions, exploring the potential of despair and betrayal in addressing ecological changes. Furthermore, HGA's critique remains within the framework of representative liberal democracy. While it critiques kleptocracy as a threat to democracy, it recurrently returns to democratic mechanisms and principles in hopes of reforming and salvaging the system from within, which can ultimately reinforce the status quo.

HGA acknowledges the role of consumer capitalism in contributing to environmental issues, but it primarily targets energy conglomerates and governments as its enablers. In critiquing capitalism and kleptocracy, HGA recognizes the symbiotic relationships between these systems and neoliberal democracy. However, it does not envision an alternative to resolve these contradictions. Mostly addressing a middle class white audience, moreover, the series does not delve deeply into class and gender dynamics under capitalism and their links to environmental degradation. It stops short of explicitly conveying that "capital organize[s] the exploitation of nature (the natural conditions of production) through the exploitation of human work" (Barca, 2024, p. 23; see also Fraser, 2022). For a critique to be holistic and effective, it needs to "see the history of labor as an important component of environmental history, noting that 'the more that (human-modified) nature is seen as the history of labor, property, exploitation, and social struggle, the greater will be the chances of a sustainable, equitable, and socially just future'" (Barca, 2024, p. 23).

Integrating labor into the equation, especially within a context marked by colonial history and land expropriation, can offer a more comprehensive understanding of how these elements intersect with emancipatory worlding. This approach acknowledges that the exploitation of both human labor and natural resources has often been intertwined, particularly in colonial contexts, and that addressing environmental degradation requires a nuanced understanding of these complex intersections. Emphasizing the role of labor highlights the potential for collective action and solidarity in striving for a sustainable and equitable future for the Exhausted of the Earth, where environmental and social emancipation are intertwined goals.

The series touches on themes of environmental justice, colonial dynamics, and, to a lesser extent, multispecies justice. However, it often falls short of providing a sustained and nuanced exploration of these topics. Its focus is largely local, aiming primarily to address issues arising within the Australian context. Nevertheless, given the colonial history of the country and the devastating impact of environmental disasters like wildfires on more-than-human life, this context holds great potential for advancing multispecies justice and even emancipatory worlding. While its critiques of capitalism, colonial worlding, and resilience are somewhat limited, HGA is notably more radical and much bolder than most television shows and cinematic productions. In the next chapter, I will demonstrate how a novel, being even less constrained by the demands of the multimodal cultural production economy, can offer an even more rigorous and cogent critique of these issues.

## References

APA. (2009). *Psychology & global climate change: Addressing a multifaceted phenomenon and set of challenges*. Retrieved April 1, 2022, from, https://www.apa.org/science/about/publications/climate-change

Barca, S. (2024). *Workers of the earth: Labour, ecology and reproduction in the age of climate change*. Pluto Press.

Bogad, L. M. (2016). *Tactical performance: The theory and practice of serious play*. Routledge.

Coulson, S. (2007). Funnier than unhappiness: Adorno and the art of laughter. *New German Critique, 100*, 141–163. http://www.jstor.org/stable/27669190

Davidson, H. (2017, August 2). Emissions from NT's McArthur basin would dwarf those from Adani coalmine. *The Guardian*. https://www.theguardian.com/environment/2017/aug/02/emissions-from-nts-mcarthur-basin-would-dwarf-those-from-adani-coalmine

Foucault, M. (1977). *Discipline and punish: The birth of the prison*. Pantheon Books.

Foucault, M. (1991). *Governmentality*. In G. Burchell, C. Gordon, & P. Miller (Eds.), *The Foucault effect: Studies in governmentality* (pp. 87–104). University of Chicago Press.

Fraser, N. (2022). *Cannibal capitalism: How our system is devouring democracy, care, and the planet—and what we can do about it*. Verso.

Garrard, G. (2023). *Ecocriticism* (3rd ed.). Routledge.

Haraway, D. (2015). Anthropocene, capitalocene, plantationocene, chthulucene: Making kin. *Environmental Humanities, 6*(1), 159–165. https://doi.org/10.1215/22011919-3615934

Haraway, D. (2016). *Staying with the trouble: Making kin in the Chthulucene*. Duke University Press.

Juice Media. (2016). Honest Government Ad | Adani Coal Mine [Video]. YouTube. https://youtu.be/n_C8S4Bz91M

Juice Media. (2017). Honest Government Ad | The Great Australian Bight [Video]. YouTube. https://youtu.be/rNUxQ-9AZss?si=oVscXxHNx9vp-fPv

Juice Media. (2018). Honest Government Ad | Renewable Energy [Video]. YouTube. https://youtu.be/sitPeRlTdNs?si=aW76QdWp28oxDeRV

Juice Media. (2019a). Honest Government Ad | Climate Emergency & School Strikes [Video]. YouTube. https://youtu.be/OL8a1YEhk_o

Juice Media. (2019b). Honest Government Ad | We're Fine [Video]. YouTube. https://youtu.be/cOmdkN6MOwU

Juice Media. (2020a). Honest Government Ad | Coronavirus: Flatten The Curve [Video]. YouTube. https://youtu.be/Hks6Nq7g6P4

Juice Media. (2020b). Honest Government Ad | Gas-led Recovery [Video]. YouTube. https://youtu.be/7CMUtKWubxg

Juice Media. (2020c). Honest Government Ad | Kyoto Carryover Credits [Video]. YouTube. https://youtu.be/92t8np88fEI

Juice Media. (2020d). Honest Government Ad | The Fires [Video]. YouTube. https://youtu.be/6BmbvTvFQ3g?si=23FPAkgLH4dBymVi

Juice Media. (2021a). Honest Government Ad | COP26 Climate Summit [Video]. YouTube. https://youtu.be/QIyKmqEdgR4

Juice Media. (2021b). Honest Government Ad | Net Zero (feat. Greta Thunberg) [Video]. YouTube. https://youtu.be/1FqXTCvDLeo

Juice Media. (2022a). Honest Government Ad | Carbon Credits & Offsets [Video]. YouTube. https://youtu.be/iCRDseUEEsg

Juice Media. (2022b). Honest Government Ad | GOV.UK [Video]. YouTube. https://youtu.be/qyt3Op2dTc0

Juice Media. (2022c). Honest Government Ad | The Supreme Court (feat. Pixelsmixel) [Video]. YouTube. https://youtu.be/jL5LaLT2BJM

Juice Media. (2022d). Honest Government Ad | Visit Brazil! [Video]. YouTube. https://youtu.be/sIdwf717p0w

Juice Media. (2022e). Honest Government Ad | Visit WA! [Video]. YouTube. https://youtu.be/Key8y1yg2yQ

Juice Media. (2023a). Honest Government Ad | COP31 [Video]. YouTube. https://youtu.be/YBjyxA-wYS4

Juice Media. (2023b). Honest Government Ad | Labor [Video]. YouTube. https://youtu.be/fHB0vDhdM3c

Juice Media. (2023c). Honest Government Ad | the Safeguard Mechanism [Video]. YouTube. https://youtu.be/YrkE_VaMD4k

Juice Media. (2023d). Honest Government Ad | Visit New South Wales! [Video]. YouTube. https://youtu.be/dcqA5Sq8Dqw

Martin, S. (2021, April 26). Scott Morrison tells Christian conference he was called to do God's work as prime minister. *The Guardian*. https://www.theguardian.com/australia-news/2021/apr/26/scott-morrison-tells-christian-conference-he-was-called-to-do-gods-work-as-prime-minister

Morrison, S. (2017). Scott Morrison Brings in Lump of Coal Into House of Representatives. [Video]. YouTube. https://youtu.be/ea5bOaPkZpc

Pini, M. (2022, July 21). Morrison courts Pentecostal world domination. *Independent Australia*. https://independentaustralia.net/politics/politics-display/morrison-courts-pentecostal-world-domination,16582

Riofrancos, T. (2025). *Extraction: The Frontiers of Green Capitalism*. W. W. Norton.

Srnicek, N. (2017). *Platform capitalism*. Polity Press.

Strick, M. (2021). How humor can promote central-route persuasion: The role of ambivalence. In M. Strick & T. E. Ford (Eds.), *The social psychology of humor* (pp. 3–19). Routledge.

Varoufakis, Y. (2023). *Technofeudalism: What killed capitalism*. Penguin.

Zekavat, M., & Scheel, T. (2023). *Satire, humor, and environmental crises* (1st ed.). Routledge.

CHAPTER 6

# Symbiotic Relations at the Intersection of Bodies of Water

Resisting and subverting deeply entrenched systems such as capitalism, neoliberalism, or the culture industry pose numerous complex questions and challenges. So does attempting to undermine anthropocentrism through anthropic means like language. These systems are intricately woven into the fabric of society and often perpetuate cycles of power and control that are difficult to dismantle.

In their critique of the culture industry, Horkheimer and Adorno (2016) propose that avant-garde literature and art serve as potent antidotes to the pervasive influences of mass-produced culture. They argue that while the culture industry tends to commodify and standardize artistic expressions, thereby dulling critical thought and individual autonomy, avant-garde works challenge conventional norms and provoke deeper reflection. In this chapter, therefore, I turn into analyzing a contemporary novel that leverages satire for envisioning alternative environmental discourses.

Art and culture play a crucial role in fostering discourses that can ultimately drive societal, economic, and political transformations. They possess the capacity to shape how we grasp, define, represent, and communicate issues like ecological mutations, bolstered by their compelling imaginative and emotional appeal. The discursive nature of literature, in particular, has prompted extensive critical examinations of the interplay between the tangible, fragile reality of nature and environment, and the discursive


M. Zekavat, *Leveraging Satire for Environmental Advocacy*,
https://doi.org/10.1007/978-3-032-04838-7_6

construction of reality. Ecocritics examining works of climate fiction, for example, elucidate how such narratives utilize language to engage our imagination and offer premonitions about the bleak future that awaits us and subsequent generations if the current trajectory of $CO_2$ emissions continues.

Less attention has been devoted to the visceral experiences that reading climate fiction can evoke (Caracciolo, 2022; Caracciolo et al., 2021; Wiek von Mossner, 2020). Notably, most climate fiction is grim and predominantly relies on fear appeal to caution its audience about a looming apocalyptic future, engaging visceral processes in the hope of prompting action. Although fear can motivate some individuals to reconsider their behavior and take environmental action, its exclusive use carries several potential drawbacks.

There are compelling arguments for exploring alternative forms of environmental narratives that extend beyond the use of apocalyptic fear. Environmental alarmism, while attention-grabbing, can ultimately prove counterproductive. Wagner (2023, p. 927) acknowledges that "environmental literature struggles to balance between apocalyptic fear at the current state of climate change and hope that there remains a chance for the survival of humans and nonhumans alike." She suggests that the environmental grotesque, with its transformative and adaptive qualities, offers a more effective approach.

Humor and satire are viable alternative, or at least complementary, strategies for environmental advocacy. These forms of expression, prevalent across cultures, possess the potential to communicate scientific findings in a manner that fosters genuine engagement and a commitment to change. Their imaginative, experiential, affective, visceral, emotional, tangible, and aesthetic appeal make them particularly effective in capturing attention and inspiring action. As Boykoff (2019) notes, humor and satire can translate complex scientific concepts into accessible and relatable narratives, thereby enhancing public understanding and encouraging proactive environmental behavior.

A relatively limited body of traditional literary works engages with ecological crises through satire. Notable authors in this niche include Margaret Atwood, Ian McEwan, Kurt Vonnegut, Don DeLillo, Philip K. Dick, and Saci Lloyd. Likewise, Douglas Adams and Mark Carwardine's *Last Chance to See*, Oliver Jeffers' *The Fate of Fausto*, and occasional pieces like Aoise Stratford's *Turtle Beach* and Italo Calvino's "Theodora" in *Invisible Cities* contribute to this genre.

Like the creative side, the situation is similarly limited in the realm of critical writings and scholarship. Despite the potential synergies between critical humor studies and environmental humanities, these fields have largely overlooked each other. Garrard (2023, p. 18) observes that "there is no shortage of popular environmental comedy, but almost no recognition of it in academic circles." While there are occasional unsubstantiated claims, such as Sørensen (2016, p. 9; see also p. 23) assertion that "humorous political stunts can also bring attention to issues that are neglected or a company profiting from environmental exploitation or human suffering," sustained academic engagement with satiric environmental critique remains relatively rare (Bevis, 2013; Branch, 2014; Chandler, 2014; DiCaglio, 2015; Grimbeek, 2017; Meeker, 1974; Peiffer, 2000; Traub, 2018; Zekavat, 2019).

This chapter responds to Michael Branch's call for "environmental writers, scholars, and activists ... [to] enter the New Age by joyfully embracing the nurturing, healing power of humor" (Branch, 2014, p. 380). It argues that Ned Beauman's novel *Venomous Lumpsucker* (2022) utilizes ambivalence and uncertainty emotions and emphasizes material agency to provide a critical reflection on contemporary political and economic issues, particularly at the intersection of technocracy and anthropogenic ecological mutations. In doing so, the chapter aligns with what Seymour (2018) describes as "bad environmentalism," which encourages the environmental movement to reconsider its association with doom and gloom by incorporating irreverence and satire.

Ned Beauman, born in 1985 in London, is an accomplished author whose debut novel *Boxer, Beetle* was published in 2010. The novel garnered significant recognition, winning both the UK Writers' Guild Award and the Goldberg Prize for Outstanding Debut Fiction. Beauman's subsequent works have also received critical acclaim and various accolades. In addition to his novels, he has contributed to several prestigious publications, including *The New York Times*, *The Wall Street Journal*, and *London Review of Books*. He has also written for BBC Radio 3 and 4. His fifth novel, *Venomous Lumpsucker*, was published in 2022 (https://www.ned-beauman.co.uk/).

*Venomous Lumpsucker* depicts the near future in which climate change and inept techno-fixes to engineer humanity's way out of it prompt an extinction industry in capitalism. The novel is set at a time when "the 2°C warming era [seems], in hindsight so gentle, so forbearing, so quaint" (Ch 2) and when Arequipa no longer features gas cold nights, rather "it's like the new Brazilian Riviera" now (Ch 3). Affluent corporations and nations are enabled to eradicate endangered species either by paying a miniscule price otherwise known as extinction credit or by capturing illusions of those species through preserving their biological specimens and genomic information in biobanks across the world.

Part of the appeal of the novel originates in its international scope spanning over Qatar, Jordan, Japan, Netherlands, Germany, Finland, Estonia, UK, and the US in the near future when the sheer mentioning of the name of US is considered to be embarrassing. When one of the characters refers to the country by its name, "everybody but her looked at the floor. The avoidance of any direct reference to that country—a custom adopted in the late 2020s out of sheer embarrassment—was these days so strictly observed that … it was genuinely startling to hear somebody say the words" (Ch 15).

The narrative highlights how the environmental decisions and behaviors of Brit Mark Halyard are influenced by broader social and ecological forces. In this sense, the novel attributes agency to nature and vibrant matter, suggesting that other species can shape the economy and power structures of Homo sapiens. This perspective challenges traditional anthropocentric views, emphasizing the interconnectedness of human and more-than-human entities in determining the future of the planet.

The novel opens with the author's note clarifying that "to minimize any need for mental arithmetic on the reader's part, sums of money are presented as if the euro has retained its 2022 value with no inflation. This is the sole respect in which the story deviates from how things will actually unfold." The author talks with such an air of confidence as if he is fully convinced that he is writing the history of the future, with only a miniscule deviation. Even before the first chapter, therefore, the prophetic and at the same time tongue-in-cheek tone starts to introduce uncertainty and ambivalence by undermining the message. Anthony Cummings describes this as embodying a "twinkling deadpan" and a "mischievous intelligence" characteristic of Beauman's writing. Consequently, the note serves as an invitation to playfully engage with the narrative as a lively satire rather than a realist account (James, 2024, p. 334).

The plot starts to unfold when Karin Resaint, a former climate skeptic who now wields her expertise in biology to save endangered species, finishes her study on behalf of the Brahmasamudram Mining Company onboard the ship Varuna and releases the fish specimens *Cyclopterus venenatus* to their habitat by using a drone that "would fly about twenty kilometers to the South Kvarken [i.e., the narrow stretch of sea between Finnish Åland and Sweden] reefs where venomous lumpsuckers gathered every breeding season, and then dump out the contents of the tank" (Ch 1). We later learn that an unfortunate technological glitch has led the autonomous mining vehicles (AMV) of the same mining corporation to mine the fish's habitat, killing the last known members of the species.

Corporate irresponsibility, callousness, and greed trigger a chain of unfortunate events and personal avarice piles up further complications. Implicated in driving a species into extinction, Brahmasamudram Mining Company has to pay with extinction credits for erasing the fish, an insignificant price for their greed and the huge profit they will subsequently make. However, Brit Mark Halyard, their 38-year-old Environmental Impact Coordinator for Northern Europe of Australia with some Thai descent, has short-sold the 13 credits the company owned to pocket the profit. His plan badly backfires when an attack on biobanks soars the price of credits many folds.

## Multispecies Symbiotic Ethics and Agency

*Venomous Lumpsucker* undertakes to offer a trenchant critique of anthropocentrism. This is achieved primarily through challenging the narratives of domination and stewardship as well as depicting the agency of natural elements.

The novel manages to challenge both narratives of domination and stewardship by creating effective ethical dilemmas. Fully endorsing their dominant position of mastery and blind to their existence as holobionts, Homo sapiens undertake to decide which species is worth saving. Using arbitrary and baseless criteria like sentience and intelligence helps them justify their decisions. Karin Resaint is hired by the mining corporation to this end: "Officially she was here on the *Varuna* to evaluate, on behalf of the Brahmasamudram Mining Company, whether the venomous lumpsucker exceeded a certain threshold of 'intelligence'" (Ch 1).

But why do human beings assume they are a good measure for determining intelligence among all species? Moreover, the concept is so vague and diluted that it cannot possibly serve as a selection criterion: it is "a word so scientifically and philosophically embattled that it was almost useless, churned to mud, but that nevertheless had implications for a company who might want to mine a species' breeding ground" (Ch 1). Human beings manipulate claims, or rather illusions, to objectivity to back their position of domination and stewardship and justify why they are in a position to determine the fates of other beings. The emptiness of these claims are exposed not only in the indecisions that Resaint and Halyard grapple with throughout the novel, but also apparent in the hypocrisy of all players and regulators as "the corporations and governments paying for these tests did not want testers who were secretly on the animals' side" (Ch 3).

At one point, Halyard remarks,

> we all have to pretend for the public that we really honestly think scanning an animal's brain and putting it in a computer is as good as keeping it alive, but there's no reason to bullshit each other behind closed doors. Nobody cares about most of these species. Nobody cares about the rusty pipistrelle or the legless skink. Even the activists don't care about them—not individually. So once the poor little fucker gets scanned, that's it. That's the end. It's just going to sit there in the database forever like some moldy old library book that nobody will ever read. It's pointless. They might as well just wipe the data after a week to save on server space, it's not like anyone would ever know the difference. (Ch 2)

The instruments that uphold the systems have no illusions about the futility and absurdity of what they do. This can also be read as a criticism of some scientists enabling and aggravating environmental destruction through false promises that only serve their own career advancements. It criticizes hypocrisy and sheer absence of integrity at different levels in the regulatory systems designed for environmental protection.

Hypocritical anthropocentrism goes hand in hand with speciesism. In the conversation that follows about Ismayilov's late beloved wife, it becomes clear that characters consider and treat Homo sapiens as morally superior and more valuable than members of other species. Halyard's colleague Ismayilov has had his wife transcribed, i.e., her brain scanned in hopes of resurrecting or cloning her after her death. "When you scanned the brain of a human being, it was because you hoped that one day you might be able to revive that human being in some new incorruptible form. But when you scanned the brain of a rusty pipistrelle or a legless skink, it was only because, after the whole species went extinct, the continuing existence of the scan would make it easier to argue in a legal or regulatory context that the species was in some sense *not yet lost*" (Ch 2). Ironically, while people try to resurrect the dead, they are inevitably causing the extinction of almost all species including Homo sapiens. The rights of all sentient beings and material agents are not appreciated and defended for their intrinsic value; as a result, all species are not consistently considered and treated ethically. This passage also takes a jab at the inefficacy of regulatory systems and miscarriages of (multispecies) justice.

Criticism of anthropocentrism goes beyond ethical dilemmas and the logic of domination (Warren, 2000). The novel strives to make reparations by distributing the agency across the ecosphere. Eschewing anthropocentrism, it refuses to attribute agency exclusively to Homo sapiens as the sole force behind regulating climate events, ecological mutations, and interspecies disruptions. Not only does it acknowledge the ecospheric determinants of environmental behavior, but it also goes a step further in delineating interspecies impact on human artifacts. One can argue, for instance, that ecospheric, rather than purely human, forces determine the global political economy in the novel. It depicts how extinction causes an economic turmoil, and how biodiversity determines the profits that each country and political system can reap from environmental degradation and extinction. On a smaller scale, it is the venomous lumpsucker that determines the fates of major characters in the novel.

In Chapter 9, we learn that, "locusts ... now ... [swarm] as far north as the Alps, billions and billions of them obliterating the wheat fields like some unthinkable weapon." Though a consequence of anthropogenic ecological mutation, this underlines ecospheric forces that drastically disrupt food security for Homo sapiens, and can subsequently cause political, social, and economic instability. In the near future, the negative ecospheric impacts will no longer be limited to the Global South that has been paying the cost of climate change in the form of stronger and more frequent famine and pest crises, despite its comparatively smaller role in causing the crisis in the first place and more restricted resources in ameliorating it.

Beaumann resorts to what Appadurai (1986, p. 5) has termed "methodological fetishism: a return to the things themselves as socially alive and active in a primary sense" (qtd. in Malafouris, 2013, p. 133) to aid in creating a discourse that can explore the mutual constitution of individuals and objects. To do so, Beaumann foregrounds material agency to escape the spell of the logic of representation. Agency and intentionality might not strike one as innate characteristics of things at first; however, one must not presume they are innate human properties either. Rather, "they are emergent properties of material engagement" (Malafouris, 2013, p. 149).

Malafouris builds on Actor-Network Theory to develop his Material Engagement Theory. Actor-Network Theory is characterized as an approach to materiality that treats human and more-than-human agents symmetrically. By conceptualizing agency as distributed and situated within relational networks of individuals and objects, Actor-Network Theory asserts that all entities involved in these networks should be analytically regarded as equally significant. In this framework, what we identify as actors or agents are fundamentally seen as products or effects of these networks. Consequently, Actor-Network Theory rejects any inherent primacy of human actors—whether individual or collective—over more-than-human actors on a priori grounds. Although this perspective may appear to reconcile the longstanding dichotomy between agency and structure in social theory, it fundamentally diverges from that aim. By incorporating material entities into sociological analysis, Actor-Network Theory does not seek to resolve this tension but rather to disregard it, pursuing what Latour (1999, pp. 16–17) refers to as a "bypassing strategy" (Malafouris, 2013, p. 123). Agency, therefore, is distributed throughout a network, residing in the connections and relationships among entities rather than being located within the entities themselves (Malafouris, 2013, p. 127). As Malafouris (2013, p. 129) puts it, "power, intentionality, and agency are not properties of the isolated person or the isolated thing; they are properties of a chain of associations."

Agency thus arises in the interactions of individuals and vibrant matter. In other words, agency is inevitably "mediated" (Malafouris, 2013, p. 221). Neither the individual nor the firearm in isolation can be held solely responsible for the act of killing. Responsibility is situated, first, in the manner in which these two agents intersect to form a new hybrid agent—the gunman—and, second, in the socio-technical network that facilitates and enables this interaction. Action entails the convergence of both human and more-than-human elements, necessitating a shared responsibility among these components. Therefore, it is untenable to maintain distinctions between human and more-than-human entities regarding agency (Malafouris, 2013, p. 221).

According to Malafouris, human cognitive processing can be understood as a *hylonoetic field*—a conceptual landscape that extends beyond the individual to encompass the extra-organismic environment and material culture. He argues that human cognitive and emotional states are fundamentally composed of elements from their surrounding material context, suggesting that our modes of thought are not merely influenced by but are actually shaped by extracranial bodily processes and material artifacts. In his words, "the human mind exists as a historically situated actuality—that is, an emergent product of complex ecological relationships and flexible incorporative forms of material engagement" (Malafouris, 2013, p. 239). In this view, material culture functions not as a straightforward cause in a mechanistic framework, but rather as a dynamic ontological force that redirects and interrelates various efficient causes, thus highlighting the intricate interplay between cognition and the material world. Accordingly, "things become agents of change and culturally orchestrated interventions, generating their own unusual evolutionary dynamics" (Malafouris, 2013, p. 246).

Instead of engaging in debates over the ontological differences between various species, or between sentient and non-sentient beings, this approach aims to dismantle "the power differentials that endow some humans with the authority to categorize others as belonging to the human species or not" (Heise, 2016, p. 198). By acknowledging multispecies and material agencies, this perspective facilitates the development of flattened ontologies. It suggests that "we might reconcile the power differentials that lie at the core of postcolonial theory, of the environmentalism of the poor, and of environmental justice and human rights advocacies with the 'flat ontologies,' the leveling of categorical differences between kinds of beings, processes, and objects that accompanies many varieties of posthumanism" (Heise, 2016, p. 166). This leveling of the ontological ground is emancipatory as it challenges traditional hierarchies and invites a more inclusive understanding of agency across different entities and systems, thus broadening the scope for equitable interactions and interventions. Such emancipation can facilitate envisioning alternative worldings and present liberating potentials by challenging entrenched power structures and expanding our understanding of agency beyond human-centered frameworks. Not only does it question traditional hierarchies but also encourages the development of new narratives and practices that emphasize interconnectedness and symbiotic interdependence.

Satire enables Beauman to advance the narrative in this direction, highlighting the far-reaching impacts of ecospheric determinants on the economy, politics, culture, society, as well as individual and collective behavior. The ecosphere significantly influences the cultural dimensions of human life, including culinary traditions. The novel vividly illustrates how anthropogenic ecological mutations have profoundly affected cuisine. Environmental degradation is mirrored in the deterioration of traditional culinary practices. As ecosystems shift and species become endangered or extinct, the ingredients that once formed the backbone of cuisines in different regions are no longer available or sustainable. This loss not only alters the physical composition of meals but also erodes their significance in cultural heritage and identity, as food is intrinsically linked to cultural expression and communal memory.

> In Ontario, the heat dried up the sap in the maple trees; in Kenya, it wilted the coffee plants and invigorated their pests; in Champagne, it broke down the acid in the grapes; in Kentucky, it cooked the bourbon in its barrels. In Campania, the mozzarella buffaloes stopped lactating for a week after every bout of strange weather; in Iberia, a fungus hollowed out the oak trees that were supposed to drop the acorns that fed the pigs; off the coast of Tasmania, the oysters' shells dissolved in the corrosive seas; and off the coast of South Australia, the bluefin tuna starved because there were no smaller fish left to prey upon. Even … Honey Golds shriveled away because there wasn't enough water in Queensland to cultivate them. (Ch 4)

These examples of degradation illustrate the profound and interconnected effects of ecological mutations on local food systems and culinary practices. As ecosystems continue to be disrupted, the rich tapestry of global culinary traditions faces collapse, highlighting the interdependencies of vibrant matter and culture.

Most of the items listed above are considered delicacies in most cultures. This, however, does not imply that the negative cultural impacts of climate change will only be felt by the upper socio-economic classes. We learn that staple nutrients and food security are similarly impacted:

> fruit and vegetables didn't taste quite as good as they used to. But if flavor had been very gradually leaking out of produce for decades, the last of it flushed away with a terrible suddenness as Halyard entered adulthood. Potatoes, carrots, beets and apples were distinguishable only by color, not by taste. Aubergines were spongy, kale was bitter, honey was as thin and tasteless as egg white. It was as if fecund Mother Earth had been replaced overnight by one of those food service companies who mostly do internment camps. There were attempts to compensate—genetic engineering, sealed biodomes, AI farmers that nurtured every sprout and youngling with the care an anxious parent gives an only child—but as was the general trend of things, even the boldest advances in technology could not keep up with the rate of collapse. (Ch 4)

Two critical points are highlighted: first, the adverse effects of ecological mutations extend to staple food items and negatively impact the middle and working classes, undermining their access to quality and sufficient nutrition. Second, it underscores the absurdity of relying on further anthropogenic interventions—such as generative AI, genetic engineering, and biodomes—to remedy problems caused by prior interventions. Instead of providing solutions, these technologies often perpetuate the cycle of environmental harm, illustrating the need for foregrounding symbiotic relationships among holobionts.

The novel is also mindful of the disproportionate impact of anthropogenic ecological mutations on different classes and the resources they have at their disposal to address them. We see Kohlmann, a billionaire, lavish on exquisite delicacies while upper middle classes gorge on Inzidernil, and lower classes are implied to have entirely suppressed the pleasure of tasting delectable victuals. Inzidernil is among the first pharmaceutical drugs entirely developed by Artificial Intelligence (AI). It was initially developed for some endocrinal intervention, but it was its side effect that captured the market (This might remind one of Ozempic, a type 2 diabetes medicine that is frequently taken for its side effect of weight reduction). Inzidernil is often taken to desensitize your judgment against the bluntness of your food intake.

While billionaires continue to indulge in the rarefied sensory pleasures of gourmet cuisine, the middle class often resorts to desensitization methods by Inzidernil. Halyard, a self-identified gourmand, particularly grapples with the dullness of flavors that once delighted his palate. A friend had given him a bottle containing 30 tablets of this unfamiliar medication, suggesting, "the next time you know you're going to have a really depressing food day, take one of these first thing in the morning" (Ch 4). This friend, much like Halyard, found that nearly every day had become a disheartening culinary experience. However, Inzidernil does not provide a genuine solution to the problem of over-processed and flavorless food; rather, it serves as a desperate attempt to numb the symptoms of a deeper ingrained issue. As Halyard notes, "the Inzidernil didn't stop him tasting what he was eating. And it didn't fool him into thinking that what he was eating was delicious. Rather, it took away his evaluative response. He no longer felt any disappointment or resentment or grief. He no longer compared the meal against his memory of what it should have tasted like. It just didn't bother him" (Ch 4).

Rather than engaging and addressing the issue, therefore, people prefer to deny or ignore it. This state of indifference or apathy echoes themes found in popular culture, such as in "The Good, the Sad and the Drugly" (20.17; 2009) episode of *The Simpsons*, where Lisa's environmental despair leads her therapist to prescribe her "ignorital," which transports her to a euphoric state of "dizzyland" where chaos appears deceptively pleasant. In this context, both characters illustrate how individuals may resort to idealistic illusions or ignore the urgency of the issue altogether to cope with overwhelming circumstances, particularly when the disparity between their power and resources and the scale of the problem they face becomes insurmountable. Beauman's narrative highlights that this experience is not merely a matter of individual capacity but is profoundly influenced by socio-economic status, underscoring how class dynamics shape one's ability to engage with and respond to ecological and environmental challenges.

## Brexit and the Pernicious Rise of the Far Right in Europe

Ecospheric determinants of behavior are entangled in the material dimension of existence, and our response to their changes is also partly informed and shaped by the material conditions of our lives. A similar dynamic also plays out in the relation of material realities with the political sphere, because as Andreas Malm and the Zetkin Collective (2021, p. 333) argue, "ideology … is profoundly material: it exists not so much inside people's heads as in what they do." *Venomous Lumpsucker* pays particular attention to the political factor that shapes environmental behavior at individual and collective levels. It projects great concerns over the rise of far-right and extremist politics in Europe. There are hints towards right-wing parties like the AfD in Germany (Ch 15), but the main focus is on the "Hermit Kingdom" or the UK after Brexit.

The novel reminisces about the Brexit referendum (2016) and how populism and right-wing politics led the country to seclusion, hence Hermit Kingdom. The initial promise of populist politicians advocating for leaving the EU was that the country is spending the money well-deserved at home in Europe. Reverting those funds to national interests was supposed to bring prosperity to people. With the advent of Brexit, however, a labor crisis has descended on the Kingdom. Instead of improving the status of the worker, it has undermined their stance in the increasingly international labor market. "By far the cheapest source of unskilled labor in northern Europe was the Hermit Kingdom's guest worker program. Farmhands … imported from the Hermit Kingdom, were paid about €600 a month for working long days on the ranches, and the majority of that went straight into the coffers of their government, leaving them with only about €200 a month to send home to their families. Nevertheless, because of the state of the Hermit Kingdom's economy, these jobs were so much in demand that you would often have to bribe an official to secure one" (Ch 8). The government betrays the working class first by false promises and hopes and then by extrapolating two thirds of their meager income. The circumstances are so dire that people cannot secure even such a degrading job without bribing authorities. The corruption and opportunism of populist politicians are censured, but voters are not spared either. They are criticized because of buying into the inflammatory polemics of those manipulating politicians. Beauman's satire targets both leavers and remainers as he discerns how "some fractions of the working class would be susceptible to fossil fascism in a mitigation crisis, as determined as anyone to hold on to what they have got" (Malm and Zetkin Collective, 2021, p. 456).

The same government that had promised better conditions to workers under the autonomy and self-determination of nationalism and parochialism is now slaving them off to other countries. Besides taking a considerable cut of their income, the Kingdom is complicit in propagating indentured labor in terrible work conditions: "By agreement between Finland and the Hermit Kingdom, the guest workers weren't allowed to move freely around the country, so after they were evacuated they were brought to the south coast to await repatriation" (Ch 8). The irony is that both the sending and receiving governments mutually agree that the working class must be subordinated and restrained. Not only will suppressive regulation of labor secure financial gain, but it will also decrease the risk of organizing dissenting movements and consequently socio-political uprisings. Class, indentured labor, and regulatory systems are effectively manipulated to keep the status quo and avoid any danger of resurgence.

The socio-political repression through the regulation and restriction of labor mobility is just one of the challenges faced by displaced and dispossessed working classes. The evolving relationship between governments and citizens, particularly workers, into a master-slave dynamic is influenced by political, economic, and environmental factors. Environmental reasons, in particular, affect not only labor market dynamics and the economy but also the health and well-being of the working class by disrupting the concept of *sumak kawsay*[1] or good living (Zekavat, 2025, Forthcoming-b).

A poignant example is the plight of Hermit Kingdom indentured laborers who suffer from zoonoses while working on distant ranches (see Rupasinghe et al., 2021). As described in a narrative depiction: "The problem was that kaptcha, like ringworm, was zoonotic, meaning it could jump from animals to humans. After you spent a week working with infected cattle, you would look in the mirror and see swellings, rashes, sometimes even plaques of fungus as thick as the crust of phlebia on the bark of an oak tree. And even after the infection was chased away with anti-fungal medication, your appearance didn't quite go back to normal: it left behind not only scarring and mottling, but also small fatty deposits under the skin which could subtly alter and deform your features" (Ch 8). The global dynamics of political economy and labor not only degrade the environment but also subject workers to inhumane conditions and numerous adverse effects of environmental degradation or what Fraser (2016, 2022) has termed exploitation and expropriation under racialized capitalism. The economic situation, coupled with displaced workers and zoonotic diseases, can leave their complexions so altered that even advanced AI recognition and surveillance systems fail to identify them.

The government is usually understood as an institution responsible for taking care of citizens and improving their life and wellbeing. The Social Contract Theory and Natural Rights Theory are foundational concepts in political philosophy, significantly influencing the development of modern democratic states. Social Contract Theory, as articulated by thinkers such as Thomas Hobbes, John Locke, and Jean-Jacques Rousseau, posits that governmental authority originates from a consensual agreement among individuals to form a society that protects their collective interests. According to Hobbes, without this agreement, life would be "solitary, poor, nasty, brutish, and short," necessitating a powerful sovereign to impose order. Locke's interpretation introduced the notion of government as a protective mechanism specifically tasked with securing individual rights, setting the stage for liberal democratic governance. Complementing the social contract, Natural Rights Theory argues that individuals inherently possess inalienable rights, including life, liberty, and property,

[1]Alternative terms include buen vivir, One Health, or planetary health (Zekavat, Forthcoming-a).

simply by virtue of their humanity. Locke advanced this notion by asserting that the legitimacy of any government rests on its commitment to safeguarding these rights, and that citizens hold the moral prerogative to resist or alter any regime that fails in this duty. Together, these theories emphasize the ethical and political imperative for governments to derive their power from the governed while prioritizing individual rights, underpinning much of the framework in contemporary human rights and constitutional law (Besussi, 2017; Estlund, 2012; Morris, 2003).

The actions of the Hermit Kingdom starkly contradict the foundational principles of governance. Rather than safeguarding the welfare of its citizens, the government exploits the labor force by collaborating with foreign entities to suppress wages and actively obstructs humanitarian aid intended for its populace. For instance, the narrative reveals that the country's navy attempted to "stop freighters full of EU humanitarian aid from reaching its shores" under the pretext of "defending the Channel" by deploying naval mines (Ch 16).

These policies, compounded by the challenges of climate change, have further exacerbated the hardships faced by the Kingdom and its people. The situation has deteriorated to such an extent that the Kingdom has descended into the status of "a country where people would fight over a job paying eight euros a day" while subsisting on "nutrient bars" (Ch 10). This degradation not only undermines the country's position in the global (dis)order but also perpetuates a vicious cycle of decline and hardship.

This scenario not only underscores the regime's blatant disregard for the welfare of its citizens but also exemplifies a broader strategy of oppression that fundamentally undermines the very tenets of Social Contract Theory and Natural Rights Theory. While these theories advocate for the protection and promotion of individual rights as the foundation of legitimate governance, the actions of the Hermit Kingdom reveal a systematic betrayal of those principles. Instead of serving the interests of the populace, the government perpetuates a cycle of exploitation and repression, ultimately reinforcing its own power at the expense of the rights and dignity of its citizens.

The novel reminisces about the time when back,

> In 2019, the Hermit Kingdom was still the sixth biggest economy in the world. So that was its official baseline. But by the time of Chiu Chiu's death, well into the process of sealing itself off completely, it had fallen to around eightieth or ninetieth biggest (there was so little data leaking out that the

> analysts could only estimate from the industrial activity captured in satellite photos). In other words, the WCSE [i.e., World Commission on Species Extinction responsible for regulating the extinction industry] treated the Hermit Kingdom like a real bruiser—not quite a Mexico, because it didn't have the same tropical biodiversity, but certainly the sort of country you'd expect to be busy draining its marshes, spraying its fields, poisoning its rivers and tarring its air, eradicating plenty of species along the way. Whereas in fact none of that was happening anymore: out of sheer poverty, it had become an environmental paragon. So a generous allocation of credits was grandfathered across to the Hermit Kingdom, almost none of which it had any use for, meaning it was free to sell them on the open market. For a while, in the early days of the WCSE, it looked as if extinction credits might be a nice boost to the Hermit Kingdom's shriveled GDP; this was thought to be why the Hermit Kingdom happily signed up to the WCSE even as it was withdrawing from every other international treaty. Then extinction credits plummeted in price, partly because the Hermit Kingdom was just one of a number of countries flogging their surplus every year, and that hope died. (Ch 10)

The novel critiques the Hermit Kingdom, highlighting its flawed policies and incompetence. However, its shrinking economy unexpectedly yields a positive outcome: the country transforms into "an environmental paragon." Advocates of various sustainability models—including circular economy, degrowth, and ecosocialism—argue that limiting economic growth is essential for genuinely addressing ecological crises. Many developing countries and former colonies remain skeptical, as they justifiably contend that developed nations have reaped immense benefits from cheap and accessible energy to fuel their growth, and now, as developing countries seek to follow a similar trajectory, they face discouragement. While the reduction in economic activity may lead to environmental gains, it poses significant challenges for developing nations striving to enhance their standard of living.

Environmental justice seeks to address this conundrum by advocating for solutions that recognize the disproportionate roles different communities have played in causing ecological crises. It emphasizes that while all communities are affected by environmental challenges, they vary significantly in their contributions to these issues and in the resources that they have available to mitigate the unfolding crises. By highlighting these disparities, environmental justice calls for equitable approaches that ensure communities most affected by environmental degradation receive the support and resources necessary to effectively address and manage these challenges.

WCSE appears to operate with the goal of ensuring that various countries receive a fair share of the extinction industry's benefits, potentially aiding in the restoration of their ecosystems. However, these efforts seem to be faltering and may even be counterproductive. One significant issue is the apparent disregard for the Kingdom's colonial and imperial history, which is a critical factor in shaping the current environmental and socioeconomic landscape. Moreover, the efforts to improve environmental conditions seem inherently contradictory. While some measures may require the reduction of economic activities to protect natural resources, the pursuit of environmental justice might simultaneously necessitate stimulating certain economies to ensure equitable access to resources and opportunities.

Ultimately, the plans put forth by WCSE inadvertently contribute to aggravating environmental degradation, particularly due to the mechanisms of market capitalism as embodied in the actions of a deranged billionaire within the Hermit Kingdom. This situation exacerbates existing injustices and disproportionately empowers the wealthy, highlighting a critical flaw in the current system. Despite an overall shrinking economy, their net negative environmental impact continues to escalate. This can be attributed to individuals like Barka, who possess substantial disposable capital and exploit their resources in ways that further harm the environment.

These twists underscore the intricacies and, I would argue, inherent flaws in the project of environmental justice, as will be further discussed in the conclusion. In the novel, environmental justice fails to halt or even slow down environmental degradation, nor does it provide tangible benefits to the Exhausted of the Earth. Instead, paradoxically, it serves to further empower billionaires, widening the wealth gap, and exacerbating environmental degradation. The narrative calls for a more effective framework, what I call emancipatory worlding, that can deliver meaningful change for both the planet and its most vulnerable inhabitants.

The displacement of vulnerable populations due to ecological mutations further illustrates how ecospheric factors can shape the political sphere, subsequently influencing individuals' environmental behaviors and choices. As ecological mutations render certain areas increasingly uninhabitable, people are forced to leave their homes. In response, some right-wing politicians and demagogues exploit this situation by stoking fear and hatred, demonizing refugees and framing them as threats (Malm and Zetkin Collective, 2021). This incendiary rhetoric affects the heartland in public attitudes and policies, impacting how issues related to climate change, immigration, and asylum are addressed in these regions. Wealthy nations in the so-called Global North often neglect their responsibilities in terms of accountability and restitution for climate-induced displacement. Instead, they manipulate climate change as a tool to reinforce xenophobic narratives, further marginalizing displaced populations and undermining efforts to address the root causes of both environmental and humanitarian crises.

*Venomous Lumpsucker* addresses the "refugee crisis" within the European Union, with a particular emphasis on "environmental refugees" (Ch 10). The novel paints a stark picture of the dire conditions within refugee camps, as explored in Chapter 9.

> It was possible, in fact, that the camp used the very same Antichain software to administer its humans that Delta Ecological Services used to administer their animals. Even after all the uproar about Kashmir, Antichain maintained a pretty dominant market share amongst the refugee camps of the developed world. Bad publicity never made much difference, not when they could do so much for a stressed-out government, transforming a "jungle" or a "tinderbox" into a perfectly transparent and predictable machine. (Just this week there had been a story about Ferenc Barka failing to turn up to a European Parliament hearing on civil liberties he'd promised to give evidence to. Nobody seemed to know where he was instead. And what was the European Parliament going to do about it? Send European Parliament gunships?) (Ch 9).

The novel chastises the dehumanization of asylum seekers in the West, attributing part of this issue to the complicity of technology, particularly AI. These technologies are depicted as tools of surveillance rather than as means to enhance the well-being of all species, including humans. Barka, a fictional billionaire reminiscent of Elon Musk, is portrayed as having amassed wealth through big tech. Despite the awareness of national governments and European legislative bodies regarding the potential dangers posed by big tech, their regulatory efforts are limited to summoning industry leaders for hearings. Barka's contemptuous absence from these hearings highlights the European Parliament's impotence in enacting and enforcing measures for the public good. This situation underscores the inefficacies of regulatory systems and the broader determinants of environmental behavior. Moreover, this passage also exposes the way metaphors are manipulated to frame, hence justify, inhuman policies and practices in foregrounding that blatant governmental and tech behavior are circumvented by framing it as a solution that can transform a jungle or a tinderbox to a machine.

*Venomous Lumpsucker* intricately weaves ecospheric and material agency, political exploitation, and the dehumanization of vulnerable populations, revealing the profound interconnectedness of these issues. The narrative critiques the Hermit Kingdom's government for engaging in practices that betray its citizens and perpetuate oppression under the guise of protecting national interests. It highlights the dangers of populist rhetoric that exploits environmental crises to scapegoat marginalized groups, while neglecting the historical and socio-economic contexts that shape these dynamics.

By illustrating the detrimental effects of environmental degradation and political manipulation, the novel calls for a nuanced understanding of sustainability that incorporates social justice and economic equity. It critiques the role of technology in exacerbating these issues, emphasizing the urgent need for comprehensive policies that prioritize the well-being of all species and challenge prevailing narratives justifying exploitation and inequality. *Venomous Lumpsucker* strives to convey that addressing climate change and social injustice requires an integrated approach acknowledging the complexities of human and ecological symbiotic interdependence. It urges readers to reconsider pathways toward a more equitable and sustainable future, advocating for systemic change that considers the intricate ties between *sumak kawsay*, socio-political well-being, and vibrant matter.

## Ecomodernism and Surveillance Capitalism

*Venomous Lumpsucker* extends the critique beyond the role of technology in surveillance to encompass a broader criticism of ecomodernism. Its critique echoes the perspective of Bonneuil and Fressoz (2016, n.p.), who caution against "the scientistic illusion that ecological awareness and 'salvation' can only come from scientists and not also from the struggles and initiatives of other Earthlings and citizens of the planet." The crucial importance of grassroots movements in addressing ecological and social challenges is underscored. It further advocates for these movements to employ a diverse array of strategies in their struggles, rather than relying solely on scientific authority. This is particularly important in an era where access to wealth and resources is often portrayed as a merit and virtue, allowing a privileged few to exploit the planet at the expense of everyone else and all species. As Ghosh (2024, n.p.) notes, "at a time when elite hucksters and all-powerful billionaires are trying to peddle the idea of solar geoengineering there is nothing more important than to remember that every one of the interconnected crises that humanity now confronts is the unintended consequence of interventions conceived of by men who believed that their superior education and privilege entitled them to override all customary, common-sense constraints."

The novel satirizes ecomodernist efforts by illustrating how these attempts can exacerbate existing challenges and by speculating on the hidden motives of their proponents. Through various examples of techno-fixes, such as spindrifters, VTOLs, AMVs, and bio-banks, it critiques the tendency of ecomodernist projects and geoengineering initiatives to overlook their potentially disastrous ecological and socio-political consequences. This critique resonates with Latour's (Latour, 2017, p. 12) observation when he writes, "in the great delirium that they call, modestly, geo-engineering, they mean to embrace the Earth as a whole. To recover from the nightmares of the past, they propose to increase still further the dosage of megalomania needed for survival in this world, which in their eyes has become a clinic for patients with frayed nerves."

The plot unfolds as an AMV mines the habitat of the venomous lumpsuckers that Resaint has released in the wake of her tests onboard Varuna.

> The *Varuna* was the home base for eight AMVs, autonomous mining vehicles, which mined ferromanganese nodules from the bottom of the Baltic. Each AMV was twenty meters long, weighed a thousand tons, and looked like a siege weapon out of *Mad Max*. At the front they had huge, spiky cutting heads to plough through the continental shelf, and at the back they had ladle chains to scoop up the nodules thus liberated. Powered by hydrogen fuel cells, they grazed the seafloor like cattle on a hillside, deciding for themselves where to go next. Of course, limits could be set in advance, and in this case the AMVs were supposed to have kept their distance from the home of the endangered fish. But something must have gone wrong. ... a team back in Mumbai were in charge of programming the AMVs, and the crew of the *Varuna* were basically just grease monkeys. (Ch 2)

Beauman satirizes the idealization of technological solutions by exposing the human shortsightedness and irresponsibility that often accompany them. Techno-fixes, such as hydrogen fuel—depicted as a solution to carbon emissions from fossil fuels—can inadvertently create new problems. Even technologies and initiatives labeled as 'green' can have detrimental environmental (let alone socio-economic) impacts if they are used to maximize the exploitation of natural resources and drive capital accumulation. The AMVs in the novel, powered by hydrogen fuel cells, exemplify this issue. While Europe has invested significant hope in hydrogen as a key component of its green energy transformation, largely due to heavy lobbying (Hamouchene, 2023), the technology is revealed to be part of a broader effort to sustain existing extractive practices and economic models.

Though these AMVs do not emit $CO_2$, their environmental impact is far from benign. Instead of promoting symbiotic multispecies flourishing, they actively destroy habitats, including those of endangered species. The narrative underscores the irony of relying on green technologies that fail to address the underlying issues of environmental exploitation and degradation. By critiquing the misplaced faith in technology as a panacea, Beauman calls for a more critical examination of how such solutions are implemented and the broader socio-political contexts in which they operate.

The common thread among various green initiatives and the excitement surrounding them is the misguided belief that any shift toward renewable energy is beneficial, and that moving away from fossil fuels is inherently positive, regardless of the approach. It is important to clarify that the climate crisis is not solely caused by fossil fuels themselves, but by their unsustainable and harmful use to power capitalist interests. To genuinely address the climate crisis—a part of the broader capitalist crisis—we must radically rethink production, distribution, consumption, and equity. Simply transitioning from fossil fuels to renewables within a capitalist framework that commodifies nature for profit will not resolve the problem. Continuing this path might worsen existing issues or create new ones related to ownership and resource exploitation (Hamouchene, 2023). Besides, as Jean-Baptiste Fressoz (2024) has convincingly shown, the notion of "energy transition"—the idea that societies progress from one energy source to another (from wood to coal, then to oil, and now to renewables)—is a misleading myth. Instead of replacing old energy sources, societies have historically added new ones on top of existing ones, leading to increased overall consumption and extractivism rather than a move toward sustainability.

Furthermore, the AMVs are deployed to explore deep ocean ferromanganese nodule fields. These nodules are viewed as potential sources of metals like manganese, cobalt, nickel, copper, and rare-earth elements, which are essential for high-tech industries (Guo et al., 2022). High-tech industries, including those promoting sustainable mobility, have been touted as key solutions to mitigate climate change, often supported by research institutes, corporations, and government policies, such as electric vehicle subsidies. While this approach has emerged as a marketing strategy to maintain existing power structures, the potential negative impacts of these technologies have largely been overlooked.

Corporations responsible for exacerbating climate change and socio-economic inequalities have found new opportunities to garner political and economic support, effectively receiving rewards. Meanwhile, there is insufficient attention to the actual environmental impact of the entire production, marketing, and distribution processes, including carbon emissions. Moreover, these technologies perpetuate social inequalities, negatively impact labor relations, and impose various burdens on many countries. This oversight leads to a cul-de-sac where further environmental degradation and socio-economic deprivation occurs under the guise of green transition, exemplified by mining activities required to supply "blood minerals" and metals for battery and electronics production (Riofrancos, 2025).

The metaphor "graze" and the simile "like cattle" used to describe AMVs ironically highlight the hypocrisy of purported efforts to combat climate change. Rather than replacing anthropocentric attitudes with a focus on interspecies symbiosis and equitable access to technology, society is replicating extractivism under the guise of sustainability by prioritizing the instrumental value of nature for producing exchange value. This approach underscores the superficial nature of some green initiatives, which continue to find new ways to exploit natural resources rather than fostering genuine ecological balance. Consequently, despite the use of sustainable fuel, the novel is set in a future where the temperature rise has far exceeded 2°C, reflecting the inadequacy of such strategies in addressing the root causes of climate change.

This passage also critiques labor dynamics and speculates on the implications of machine autonomy. The novel expresses skepticism toward machine autonomy, as evidenced by the destructive actions of the AMVs. Despite using sustainable fuel, these autonomous vehicles are depicted causing significant environmental damage. This is compounded by past incidents, such as when "a Malaysian company's rogue AMV actually had demolished a number of Bajau refugee stilthouses off the coast of Borneo" (Ch 2), highlighting a failure to learn from previous mistakes. The narrative links automation and its impact on the labor market with ecological and social injustices. By depicting an IT team in Mumbai controlling 1000-ton autonomous mining vehicles located thousands of kilometers away, the passage illustrates the catastrophic outcomes that can arise from combining automation and outsourcing—strategies corporations often use to undermine labor, cut costs, and boost profits.

The unwarranted exuberance surrounding technological breakthroughs contributes to a shortsighted embrace of techno-fixes. Spindrifters serve as yet another example of such hasty solutions that rely on geo-engineering to address climate change. Initially welcomed with enthusiasm, these devices soon reveal significant flaws. While they generate strange low-altitude storms that affect only seabirds, they also disrupt rainfall patterns over inexplicably vast distances. Given that rainfall patterns have already been severely disrupted, the introduction of such technology exacerbates existing environmental challenges: "They whisked up these eldritch low-altitude storms, which were of no concern to anyone but seabirds; but they also seemed to interfere with rainfall patterns, even at quite unaccountable distances away. And rainfall patterns had been brutalized enough already" (Ch 1).

Soon after the initial excitement, the optimism surrounding the spindrifters dissipated, much like mist, as attention shifted to new technological possibilities, leaving the fleet largely unlaunched. Nonetheless, several companies had already built the first spindrifters—a testament to the fierce competition in the race to save the world. Some of these companies collapsed before testing their prototypes, resulting in about a dozen spindrifters left adrift in the Baltic. These unmanned, wind-powered, self-navigating ghost ships, made from nearly indestructible polymers, continue their silent journeys until a mechanical failure occurs, which could take decades. Rather than regenerating biodiversity, these inventions further damage the ecosystem: "Such were the new fauna of this poisoned sea. No ringed seals anymore, no harbor porpoises, no velvet scoters, no European eels, no angel sharks, and practically no venomous lumpsuckers. But a thriving ecosystem of these faceless pack-beasts: cargo drones and spindrifters and the autonomous mining vehicles that browsed the ocean floor for ferromanganese nodules forty fathoms beneath their mothership *the Varuna*" (Ch 1).

The novel presents other instances of bioengineering interventions that similarly fail to address the root issues. A particularly absurd example is the willingness of scientists, policymakers, and corporations to redefine the concept of extinction to maintain the status quo. In this world, "even if a species didn't have a single living individual left on earth, it should still not be considered extinct, as long as it had been the subject of what was called 'multimodal preservation.'" This preservation includes DNA sequences, microbiota profiles, MRI scans of the body, connectome scans of the brain, recordings of behavior in the wild, and descriptions of habitat and diet—all stored in servers managed by various biobanks. This technological approach does not preserve endangered species; instead, it maintains their illusions. Such a shortsighted perspective fails to grasp that "a species was not just a set of bodies. It was a set of habits, relationships, territories, entanglements" (Ch 2). By relying on biobank preservation, these species are effectively reduced to mere data, stripping them of their ecological and relational contexts and material existence.

In the novel, technology is not solely depicted as a tool for bio- and geoengineering solutions but is also portrayed as a means for governments and corporations to enhance surveillance and control over society. The narrative satirizes the extensive use of surveillance technologies by these entities, showcasing a variety of tools such as scullions and drones (Ch 9), satellite tracking and applications designed to blank out cleavages (Ch 10), smart apartments (Ch 15), and artificial intelligence.

The manipulation of technology by corporations serves to reinforce their exploitation of labor and enhance surveillance capabilities. An example of this is the Brahmasamudram Mining Company's decision to station Resaint in a lab aboard the Varuna, rather than allowing her to work from the more convenient Swedish coast. This move is a psychological tactic, akin to a tribal ritual, that infiltrates even the most impersonal transactions of multinational corporations. The intention is clear: while under contract, Resaint is meant to feel she belongs to the company, confined within the boundaries of their domain, with only the cold Baltic waters surrounding her. The corporations do not hesitate to use both physical and psychological strategies to assert control over their employees.

Despite the underlying reality, the expectation is that these dynamics remain unspoken. The employee is acutely aware of her dependence and the surveillance she is under, understanding that, like all crew members, she is essentially a subordinate of the Varuna. However, her role is presented as that of an objective scientist, tasked with making impartial judgments free from external influence. This facade is crucial for all parties involved, as it preserves her "immaculate priestly aura" and legitimizes her position (Ch 1).

The lordship and bondage rapport consolidated by the surveillance system enables the corporation to influence her expert opinion. Through surveillance, Halyard becomes aware that Resaint plans to recertify the venomous lumpsucker as intelligent, a classification that would require the payment of extinction credits for mining its habitat. Anticipating this, Halyard, who has short-sold the credits prior to their price spike following an attack on biobanks, leverages this information to his advantage. He instructs Devi, the captain of Varuna, to disconnect Resaint from the network, thus preventing her from making the recertification and protecting his financial interests.

Satire is twofold in this episode. First, it critiques the way technology is exploited to advance corporate greed, optimizing profits at the expense of ethical considerations and environmental sustainability. Surveillance is used not only to monitor employee behavior but also to manipulate expert opinions to align with corporate interests, thus placing financial gain above scientific integrity. Second, it highlights the ingrained anthropocentric nature of these power dynamics, where humans manipulate regimes of truth to dictate the criteria for intelligence and determine which species are deemed sentient or worthy of protection, thereby consolidating an epistemological monopoly. This selective designation underscores a broader critique of societal tendencies to privilege certain forms of life, while exploiting and disregarding others, reflecting the flawed priorities present in corporate practices and environmental governance. Together, these elements create a compelling commentary on the intersection of sustainability, technology, corporate power, and ethical responsibility in contemporary society. This critique is further emphasized in a conversation about value, where Resaint questions, "you really don't believe that anything can have a value of its own beyond what function it serves for human beings?" (Ch 12).

Not only corporations but also governments utilize technological advancements to protect their vested interests. The novel introduces "antichain" technology, which offers pervasive surveillance capabilities. Unlike the metaphor of eyes, which suggests a limited transmission between the inner and outer worlds, antichain is more aptly described as "an all-permeating brain, and reality itself encompassed within that brain" (Ch 15). This description underscores the comprehensive and intrusive nature of such surveillance technologies, suggesting a profound integration with reality that extends beyond mere observation to encompass a deeper level of control and influence through simulacra.

The narrative reveals that antichain's capabilities extend well beyond the data they are officially contracted to manage. They have access to a broad array of proprietary data they can license, as well as public data they can scrape. This vast expanse of information allows antichain's algorithms to link any two indicators, provided that such a connection can yield a new insight (Ch 15). This comprehensive data integration grants governments near-absolute control over the lives of their citizens, significantly complicating, if not entirely impeding, any attempts at transgression or subversion. The depiction underscores the challenges posed by advanced surveillance technologies, which consolidate power and inhibit dissent.

On the other hand, individuals seek innovative ways to circumvent oppressive surveillance, with seasteading emerging as one such strategy. Initially conceived as a technological solution to help wealthy nations adapt to climate change and sea level rise, seasteading ventures have largely been deemed unsuccessful. The initial belief was that seasteads would be located in tropical regions with consistent sunny weather. However, the first substantial seasteading community, Aeolia, situated in the Atimaono Lagoon in Tahiti and established with the French Polynesian government's support, was destroyed by Cyclone Josese. The following year, another major initiative, Mainspring 1 in the Gulf of Mexico, backed by sponsors like Ferenc Barka, was seized by the Los Zetas cartel. As a result, seasteading lost its viability as a climate change solution.

Despite these setbacks, seasteading retained popularity due to its potential as a refuge from governmental surveillance. Surface Wave is one such example, floating "ten kilometers off the coast, where Vyborg Bay opened out into the gulf" (Ch 12). Entry to Surface Wave is by invitation only, and it exists beyond the jurisdiction of any nation-state, providing its residents with genuine privacy beyond the restrictive reach of surveillance. This autonomy attracts individuals seeking to escape "arbitrary and capricious governments" (Ch 13) in their quest for spaces of freedom and self-determination amidst pervasive state control.

Thus, *Venomous Lumpsucker* presents a multifaceted and nuanced critique of the complex relationships between technology, corporate interests, and environmental governance. It questions the ecomodernist belief that technological advancements can successfully separate economic growth from environmental degradation. Instead, the novel highlights how such techno-fixes often aggravate existing issues. Ecomodernism is depicted as inherently anthropocentric and technocentric, focusing on humanity's supposed responsibility to stabilize and protect nature through the intensified exploitation of resources. While this philosophy appears progressive, it is criticized for perpetuating extractivist practices under the pretense of sustainability.

Furthermore, the novel underscores the moral implications of surveillance technologies, utilized by both corporations and governments to exert control over individuals and labor. The manipulation of expert opinions and the coercive nature of corporate surveillance illustrate how technological solutions can be harnessed for corporate greed rather than communal and interspecies wellbeing. The novel's portrayal of seasteading as a failed, yet enticing, escape from government oversight highlights a desperate quest for autonomy in an increasingly monitored world, exposing the limitations and contradictions inherent in ecomodernism.

*Venomous Lumpsucker* invites readers to grapple with the complexities of technological solutions and their socio-political consequences, advocating for a more holistic approach that prioritizes ecological integrity and social justice over mere technological advancement. In doing so, it challenges the notion that progress can be achieved and evaluated without addressing the intricate relationships between technology, different communities, and the ecosphere.

## Market Capitalism and Carbon Sequestration

The novel acknowledges more-than-human agency while also highlighting the societal and structural constraints that influence individual behavior. One of the most prominent targets of its satire is capitalism and the inequalities it perpetuates. Environmental solutions rooted in globalization and market capitalism risk reinforcing existing power dynamics, leading to further disenfranchisement and dispossession. By doing so, they deepen structural inequalities rather than addressing or alleviating them.

Beauman uses market regulatory bodies as a starting point for his critique. The reforms introduced by the World Commission on Species Extinction (WCSE) have profound implications for corporate behavior, particularly through the manipulation of credit pricing. However, the system is fraught with inherent flaws, such as a lack of transparency and structural loopholes that inevitably foster corruption. The presence of privileged access to information, exemplified by actors like Smawl, and the sway of lobbyists further complicate the picture. Halyard, for example, capitalizes on insider knowledge to short sell 13 credits when they are valued at €67,000 each. His scheme only unravels when another policy failure occurs with the collapse of biobanks. These reflect how regulatory bodies can influence market behavior not only through pricing strategies but also by altering working definitions—such as redefining what constitutes extinction and preservation.

In addition to issues of corruption and transparency, the regulating authorities seem to prioritize market dynamics over genuine conservation efforts. A notable example is their decision to redefine extinction to stabilize market fluctuations. Consequently, the focus shifts from preserving actual members of a species to saving their genomic profiles and MRI scans. As an example of a shifting baseline, WCSE's capitulation to the market highlights the structural inequalities embedded in the system, designed to protect the interests of a specific class. The novel raises critical questions about the ethical and practical implications of such regulatory practices, challenging the effectiveness of market-driven solutions in addressing ecological mutations.

The introduction of genomic profiling as a means to combat biodiversity loss represents yet another attempt to reconcile conservation efforts with the demands of a capitalist economy. However, as demonstrated in the previous section, such technological fixes can, and often do, have unintended consequences. Rather than assisting conservation, these efforts can cause further complications. This is vividly illustrated by a coordinated attack on six major biobanks in Lausanne, Spitsbergen, Cape Town, Mexico City, Taizhou, and Yokohama. During these attacks, a computer worm infiltrates each facility and initiates a false chemical hazard alert, effectively evacuating staff from the cold storage areas. It then locks the doors, disables the freezers, and increases the temperature, transforming the storage units into ovens. The worm also accesses thousands of cloud servers simultaneously, erasing every file ever uploaded by the biobanks. This included critical data such as DNA sequences, connectome scans of the brain, and MRI scans of the body (Ch 4).

The aftermath of these attacks led to a significant increase in the price of extinction credits, jumping from €38,432 to €287,057. This dramatic surge prompts Halyard to embark on a desperate mission to find the venomous lumpsucker and demonstrate that it is not extinct. His journey begins at Sanctuary North in Estonia, a site that functions as both an ark and an Eden, spanning approximately 45 $km^2$ along the western shore of Lake Peipus. Sanctuary North is managed by Delta Ecological Services, which arose from the partial privatization of Estonia's Peipsiveere Nature Reserve. Initially, Delta specialized in protecting and restoring wetlands and waterways to help companies comply with the EU Water Framework Directive. However, with the advent of WCSE, Delta identified a new business model surpassing their earlier environmental initiatives. This model, grounded in market capitalism, suggested that "the job of saving species would be efficiently allocated to whomever could perform it most cheaply, achieving the same result at a minimum cost to the world economy" (Ch 6).

This approach exemplifies the paradox of using mechanisms that have historically contributed to environmental degradation as tools for conservation. Governments worldwide, including those in Estonia, Honduras, and Myanmar, fall short of their responsibilities to safeguard Earth's critical zones. Instead, they auction off invaluable ecological resources to multinational corporations, which exploit these opportunities to perpetuate their harmful practices under the guise of preservation. Delta Ecological Services serves as a prime example, prioritizing profit maximization over ecological integrity.

Employing advanced technology in landscape design and surveillance, Sanctuary North operates as a panopticon state reminiscent of Foucault's argument (1975). Within its boundaries, cameras and sensors meticulously monitor every animal, feeding data into antichain predictive analytics software. This system is designed to maintain a delicate ecological balance, intervening as needed to prevent conflicts among different species. However, fluctuations in credit prices have prompted Delta to reduce costs, including cutting back on essential system maintenance. This has resulted in a critical lack of awareness regarding the precise locations of animals within the sanctuary, undermining the sanctuary's ability to fulfill its conservation goals effectively.

Initially, the sanctuary confines its inhabitants within controlled environments, using enclosures designed to mimic their natural habitats. However, as these spaces prove inadequate, efforts are made to mitigate and manage interactions between species. Despite these interventions, the system ultimately fails to maintain the intricate balance essential for their survival. This failure highlights the inherent limitations and contradictions of attempting to engineer natural ecosystems through technological means, further exposing the challenges of aligning capitalist motives with authentic conservation efforts. The narrative underscores the difficulty of reconciling the drive for profit with the complexities of ecological preservation, emphasizing the need for approaches that prioritize ecological integrity over technological and economic expediencies.

Furthermore, Delta has resorted to depositing hazardous chemical waste on their lands as a means of supplementing their income, effectively transforming the reserve into a chemical waste dump. Despite the resulting environmental devastation, Delta remains unconcerned about potential legal repercussions, relying on blanket exemptions provided in their government agreements to ensure operational flexibility. Besides the consequences of deregulation, the company's actions expose the darker side of green technologies, as the hazardous wastes originate from a factory that produces antifouling resins for ostensibly promising green tech initiatives.

> Our agreement with the government gives us certain blanket exemptions—you know, so our operation can be agile. We cleared everything with our lawyers in advance. Also, you know where it comes from? The factory in Lyudinovo makes antifouling resins for—have you heard about reverse electrodialysis energy capture? Very promising green tech—so it's absolutely an environmental positive, but the problem is there aren't a lot of places they can go with this stuff anymore, it's not nearly as straightforward to just ship it to Africa as it would have been even 5 or 10 years ago, so they paid us a terrific rate. It was the right move at the time, even if not everyone was happy about it. And honestly I thought it would just be temporary until ZymoD could clean it up for us. (Ch 6)

The novel therefore underscores the inherent inefficacy and profit-driven motives that pervade the extinction industry, questioning its genuine contribution to conservation efforts. The Delta officer's justification for their actions, referencing "blanket exemptions" provided by the government to ensure operational agility, exposes how neoliberalism values economic flexibility over environmental responsibility. This reveals a troubling prioritization where regulatory frameworks are manipulated to serve corporate interests rather than ecological well-being. Furthermore, by presenting their activities as connected to "very promising green tech," the officer perpetuates the illusion of technofixes as viable solutions, while illustrating their failure to address the fundamental causes of ecological degradation. This narrative exposes how technological advancements can often mask underlying issues and allow harmful practices to continue under the guise of environmental progress.

Moreover, the mention of the increased difficulty in disposing of waste materials—compared to the past ease of "just ship[ping] it to

Africa"—highlights the historical exploitation of the Global South, which disproportionately suffers from environmental harm. This statement reveals a broader pattern of complicity between governments and the private sector in establishing and perpetuating sacrifice zones, regions where environmental and human health are consistently sacrificed for economic profit. The narrative critiques the superficial alignment of green technologies with sustainability, exposing how profit-driven motives frequently undermine multispecies symbiotic relationships. By drawing attention to these exploitative practices, the novel challenges the prevailing narrative that green technologies inherently contribute to sustainability, instead revealing how they can perpetuate existing inequalities and environmental injustices when primarily guided by financial incentives.

The narrative unequivocally satirizes the inadequacy of solutions rooted in racial and market capitalism for addressing environmental crises, particularly through its depiction of the extinction industry. It portrays a system where companies like Brahmasamudram Mining can essentially purchase the right to eradicate a species by spending an extinction credit. These credits permit the destruction of any species unless it is classified as "intelligent" by experts, in which case 13 credits are required. This distinction, embedded in the bureaucratic framework of WCSE, reflects an effort to economically disincentivize the extinction of intelligent species without imposing an outright ban, as such a prohibition would conflict with free market principles.

Extinction credits, allocated annually by WCSE, are distributed through a combination of free allocation and auctions. The framework is structured to gradually reduce the supply of credits, with the theoretical goal of driving up prices to a prohibitive level, thereby incentivizing innovation aimed at preventing species extinction. This laissez-faire approach relies on market despotism and the assumption that the market will self-regulate through the dynamics of supply, demand, pricing, and innovation.

However, the system is beset by the same inequalities that characterize every financial market. Lobbyists, particularly those from firms like Kohlmann Treborg Nham, have successfully integrated numerous allowances, indulgences, exceptions, and delays into the WCSE framework. Consequently, the intended scarcity of extinction credits has never materialized. Instead, these credits remain plentiful and inexpensive, readily accessible to those who can afford them. This dynamic ensures that extractivism continues to be economically viable, with credits priced at a relatively trivial €38,432 for lucrative undersea mining operations (Ch 2).

Building on its critique of inequities, the novel satirizes contemporary misguided attempts at environmental justice. One such example is the character of Selim, the deputy director of the Ministry of Forest and Water Affairs at the Regional Directorate for Southeastern Anatolia, Türkiye, whom Halyard and Resaint meet on their journey to Surface Wave. Selim elucidates the intricacies of WCSE rules, which permit Türkiye to construct a 600-kilometer rail line without incurring penalties, as it aligns with the country's baseline level of economic development. Türkiye is incentivized to avoid such developments by receiving extinction credits for each species saved, which can be used domestically or sold on the open market.

Being based on the mechanisms of market capitalism, this system transforms environmental justice into a transactional affair, prompting Selim to "commission" new endangered species for financial gain. He describes how a company can manipulate DNA to create new species, thereby earning credits for their ostensible conservation. This loophole satirizes environmental justice initiatives with their propensity to avoid complexity that can consequently give rise to manipulation. Similar scenarios unfold in other countries like Greenland and Mexico, where different baselines lead to incentives for false claims of species conservation based solely on genetic markers.

The narrative critiques these market-based solutions and the concept of market despotism by highlighting their dependence on the very capitalist logic that initially contributed to the environmental crisis. Capitalism, which inherently thrives on inequality, is ill-suited to deliver genuine justice through its mechanisms. Satire also targets the deficient system that lacks a criminal justice apparatus, relying solely on technical safeguards to prevent overtly criminal acts. By doing so, the novel underscores the inadequacy of capitalist frameworks in addressing environmental and social justice issues, suggesting that true justice requires systemic change beyond market-driven approaches.

In a pointed critique, the novel echoes a sentiment often voiced by progressives: "Leftists sometimes asserted that within a capitalist framework there could never be a solution to the extinction crisis that was untainted by profiteering and abuse, because the free market was like some malevolent AI, infinitely more devious than the humans who thought they could constrain it" (Ch 13). This reflects the belief that capitalism's inherent profit-driven motives prevent it from addressing the extinction crisis genuinely. For emancipatory worlding to emerge first we need the existing structures to collapse. To quote Salman Rushdie in *The Satanic Verses* (Rushdie, 1988, p. 3, 15, 31, 84, 182, 403), "to be born again, first you have to die."

Ultimately, the novel critiques reliance on market dynamics as a solution to environmental issues, demonstrating how these mechanisms perpetuate existing inequalities and fail to curb destructive practices. It urges a reevaluation of systems that prioritize economic interests over genuine ecological sustainability and social justice. By satirizing the idea that capitalist frameworks can effectively protect biodiversity, the narrative emphasizes the need for alternative solutions to pressing environmental challenges.

In its critique of market capitalism, the narrative segways into another troubling aspect of capitalism's intersection with environmental issues: race and indentured labor. The novel's international perspective, reflected in its diverse settings and characters, highlights the convergence of indentured labor and environmental justice under the framework of racial capitalism (see Robinson, 2019). This is exemplified by the Tinkanen migrant labor camp on the south coast of Finland, where the exploitation of marginalized workers underscores the broader systemic inequities perpetuated by capitalist structures.

The narrative highlights how globalization has morphed into contemporary forms of slavery, significantly impacting working classes and indentured laborers, particularly in relation to environmental disasters (Barca, 2020, 2024). A striking example is the spread of kaptcha, a fungal infection affecting both cattle and laborers. In central and northern Finland, milder winters have facilitated the expansion of the ranching industry, which now includes 20 million low-emission cattle in the lakelands. Despite advances in automation, the kaptcha infection has compromised biometric systems used for cattle monitoring, as the fungus alters facial features, making them unrecognizable to cameras. This system failure allows the disease to spread unchecked, further exacerbating the conditions of laborers who have relocated from the Hermit Kingdom to Finland in search of livelihood. The narrative also introduces the "rain of gnats," which originates from Surface Wave, adding to the burdens faced by indentured laborers on Finnish ranches. This development, combined with the challenges posed by kaptcha, underscores how cascading environmental challenges disproportionately affect vulnerable and marginalized populations.

These experiences culminate in a profound realization for Halyard: "He understood that the extinction industry was a great loathsome beast, and a trophy wine like that Dauvissat, served at a trophy dinner like this one, was just one of that beast's glandular fluids" (Epilogue 1). This insight drives him to infiltrate the executive class and form an alliance with a powerful figure in the lobbying sphere. His objective is to subvert the extinction industry from within, utilizing a bio-invention created by the now radicalized Warkentin. This strategic decision underscores his commitment to enacting change by challenging the system internally, using innovation as a means of disruption and transformation.

Halyard soon realizes that dismantling the extinction industry from within is more complex than he initially anticipated, particularly when faced with the formidable influence of figures like Kohlmann and Barka. Kohlmann's story exemplifies his extraordinary power and reach:

> "After all, Kohlmann was one of the founding partners of Kohlmann Treborg Nham. There was no subsection or subclause that wouldn't yield to his touch. No matter whether this regulation originated with the Belgian government or the European Union, he was probably no more than two phone calls away from the person who drafted it. And even if he couldn't arrange an emergency legislative session to get the impediment off the books before the omelette course, there was always some other way. Maybe he could lobby shrimp to stop harboring so many microbes. Maybe he could lobby God to repeal human frailty. Even a billionaire, who would try to throw money at the problem—threaten to buy out the hotel, perhaps—could only break the rules. But Kohlmann could change them" (Epilogue 1).

The immense power wielded by individuals like Kohlmann enables them not only to navigate but also to reshape the regulatory landscape to their advantage. Halyard comes to understand that the challenge lies not just in confronting the industry itself, but in grappling with the entrenched networks of influence and power that sustain it. His realization reflects on the possibility of ever changing systems entrenched with vested interests and systemic injustices.

The story of Barka similarly highlights the significant influence exerted by individuals like him within a capitalist framework, showcasing difficulties in achieving systemic change. Several months earlier, the Secretary of State for Environment, Food and Rural Affairs of the Hermit Kingdom, known as the Mermaid, uncovered a plan to sell a substantial portion of the South West Peninsula to Ferenc Barka, a prominent figure in antichain infrastructure, for the creation of a nature reserve. Barka represents a particular class of young billionaires, distinguished by a plethora of ventures and projects that often elicit a mix of boredom and distaste due to their overwhelming number and the egoistic ambition driving them. This jab satirizes the entrenched power dynamics within capitalist systems, where individuals like Barka can leverage their wealth and influence to pursue personal ambitions, often at the expense of broader societal or environmental considerations.

Despite discovering the questionable deal, the Mermaid chooses to remain silent rather than exposing Barka. Nevertheless, she quickly finds herself under surveillance, marked as a potential traitor. It seems that Barka's influence extends to providing operational support to his political allies, utilizing algorithms with near-telepathic capabilities to maintain his sway and control. Barka's charismatic persona and extravagant hobbies, such as his collection of submarines, captivate admirers like Halyard, who tend to overlook his egotism and morally ambiguous activities (Ch 15).

Barka's audacious pursuits reach a grotesque peak in a creation reminiscent of Jonathan Swift's *Gulliver's Travels*: a massive, hovering jellyfish-like structure equipped with photovoltaic skin and long, dangling cables, suspended above a dense forest. This fantastical and imposing edifice stands as a testament to Barka's immense power and resources, provoking questions about its purpose and the broader implications of such a project (Ch 16). The pervasive reach of billionaires like Barka in capitalist societies coupled with excesses of their unchecked ambition and wealth enable them to wield their influence and resources to manipulate systems to their advantage while maintaining a veneer of charisma and respectability. It raises critical questions about the societal and environmental impacts of allowing such concentrated power and wealth and the potential consequences of prioritizing personal ambition over collective well-being.

## Radical Resistance

Compared to the works analyzed in previous chapters, therefore, *Venomous Lumpsucker* offers a more nuanced and intersectional critique of capitalism, highlighting the roles of deranged billionaires, compromised politicians, labor, and the complexities of environmental and social justice. It moves beyond merely appealing to market and neoliberal solutions, instead seeking to undermine the system rather than reinforcing it. In a manner reminiscent of Chaudhury's implications in foreseeing a civil war, Beauman also entertains the notion of violence as a means of disrupting and potentially dismantling the existing system as a counteracting measure.[2] This exploration of systemic critique through unconventional means underscores the novel's willingness to challenge the status quo and confront the entrenched power structures that perpetuate inequality and environmental degradation.

The theme of resorting to violence as a means of overthrowing entrenched systems appears multiple times throughout the novel. One instance involves the Mermaid plotting to assassinate Barka, a plan that Resaint urges Halyard to thwart, arguing that Barka's actions have inadvertently led to the preservation of numerous animal species. However, it is soon revealed that Barka has gathered thousands of species in an enclosure for the sole thrill of hunting them. His rationale, influenced by a distorted interpretation of Alvin Ip's work, is based on the belief that destroying information—such as species—results in the creation of new information elsewhere. Barka expresses his exhilaration in having a significant impact on the "informational topology of the universe," describing it as "humbling to realize that I will make more of an impact on the informational topology of the universe than any other human being has ever made before me" (Ch 16).

Barka's involvement in attacks on biobanks, executed through antichain's deep learning project, X5, further underscores his manipulation of technological advancements and economic systems for personal gain. Originally intended to support the extinction credit system, biobanks have become integral to the burgeoning extinction industry. Barka therefore does not hesitate to resort to epistemological and physical violence to expedite the collapse of the system that has given rise to him. Most significantly, his explanations, though seemingly deranged, hint at the emergence of technofeudalism—a concept explored by Yanis Varoufakis in *Technofeudalism: What Killed Capitalism* (2023)—where a few individuals wield technological control and economic dominance, reshaping traditional capitalist structures. Ironically, the few that have benefitted most from the system seem to be more willing (and capable) to undermine it than the many it disenfranchises.

[2] I am putting the final touches on this chapter in the wake of Luigi Nicholas Mangione's arrest as the suspect in the killing of Brian Thompson, CEO of UnitedHealthcare, and the backlash against Elon Musk's unpopular DOGE promotion, which led to violence and vandalism against Tesla dealers and vehicles.

Radical resistance strategies resurface in the novel's critique of capitalism. Resaint's proposal—assigning the wealthiest individuals on earth a vulnerable species and threatening execution if the species goes extinct—satirically underscores the extreme measures that might be considered necessary to address inequitable socio-economic hierarchies. Her suggestion of violent resistance is portrayed as a potential means to address and somewhat flatten the entrenched socio-economic hierarchies and power structures. In a world where wealth and power are concentrated in the hands of a few, leaving the majority dispossessed, the narrative suggests that a strong government and robust laws—willing to employ acts of violence as in executions—might be necessary to restrain the powerful from their relentless exploitation of resources. This proposal highlights the desperation and urgency felt by the Exhausted of the Earth in the face of systemic inequities and environmental crises.

The theme of violence recurs at Sanctuary North, where an armed attacker confronts the protagonists. The attacker bluntly asserts, "the extinction industry has never saved a single species. It's just a performance, a fiction. It's about extracting subsidies and kickbacks, year after year. That's all. The price of credits goes up, you make money. The price of credits goes down, you make money. The suits always win and the animals always lose. A hundred thousand extinctions a year and you're just making it easier for them" (Ch 6). This confrontation emphasizes the systemic flaws and profiteering within the extinction industry, reinforcing the novel's critique of market capitalism. Violence is again suggested to be a viable and potentially effective strategy in disrupting the system that perpetuates ingrained injustices, particularly for the Exhausted of the Earth.

## Conclusion

I tried to demonstrate how satire allows Beauman in *Venomous Lumpsucker* to intricately examine the complex interplay between marine life, human societies, and socio-political institutions. The protagonist's global journey sheds light on pressing environmental issues like extinction, pollution, and ecological mutations across racial, colonial, class, and species lines (Cf. Fraser, 2022). James (2024, p. 334) observes that Beauman's novel deliberately avoids presenting its scientific accuracy through paratexts that emphasize the reality of climate change. By portraying a skeptic who adopts pro-environmental behaviors due to a personal connection with an object of care, the novel suggests that there are various ways to inspire readers toward healthier and more equitable futures, and different paths may resonate more or less with different readers. Expanding our understanding of climate fiction, its narrative resources, and its potential audiences could enrich our visceral and cognitive comprehension of the scope and urgency of climate change.

*Venomous Lumpsucker* offers a nuanced critique of the intertwined relationships among human actions, socio-political institutions, and ecological mutations. The novel targets racial and market capitalism, as well as the absurdity of billionaire-driven projects, highlighting how these systems frequently perpetuate existing inequalities. By weaving together themes of environmental justice, technological interventions, and capitalist exploitation, Beauman's work satirizes the technocratic reliance on market mechanisms and technological fixes to address problems they have historically initiated and aggravated.

Firstly, the novel critiques the notion that the best way to conserve the planet is by assigning it a price, allowing the market—considered by some as cognitively superior to public action and intervention—to integrate and self-regulate the value of nature. This idea, rooted in the belief that only private property can ensure proper management of natural resources and commoning is bound to end in tragedy (Cf. Hardin, 1968), suggests that the biosphere should be enclosed by attributing property rights to all elements and ecological functions of the Earth system. Nature, thus transformed into natural capital, becomes interchangeable with financial capital. All services provided by the Earth system, such as carbon capture, pollination, and water purification, can be monetized and traded in environmental service markets, rewarding property owners who manage these spaces. This perspective blurs the distinction between natural wealth and social value, turning nature into "the biggest business in the world," an economic entity producing value independently of human labor (Bonneuil & Fressoz, 2016; see also Battistoni, 2025).

In this framework, there is no limit to growth; environmental preservation, crises, and resource scarcity are framed as economic opportunities. In the context of neoliberalism and globalization, as promoted by WTO agreements and the financialization of the economy, environmental norms in wealthy countries have often resulted in the relocation of polluting activities to poorer nations, rather than leading to global environmental improvements (Bonneuil & Fressoz, 2016). Market capitalism is rife with contradictions and limitations in addressing anthropogenic ecological mutations as it often exacerbates existing inequalities and shifts burdens onto less affluent regions.

Secondly, the novel critiques racial capitalism by highlighting how markets for carbon credits and biodiversity credits have spurred a global movement of land appropriation. These markets, while ostensibly aimed at rewarding environmentally beneficial activities, often result in the dispossession of rural and indigenous populations from their commons (Bonneuil & Fressoz, 2016). The narrative underscores how contemporary mechanisms of environmental capitalism continue to perpetuate historical injustices, prioritizing profit over the rights and livelihoods of marginalized communities.

The novel acknowledges that capitalist and industrial modernity is constructed on inequalities related to class, race, and species—master relations that are not innate but have emerged from historical struggles and are continually contested by marginalized groups worldwide. It suggests that capitalist and industrial modernity is neither the only nor the optimal way to organize social and ecological relationships in the contemporary world (Barca, 2020, 2024).

By highlighting the intersections of market and racial capitalism with exponential anthropogenic ecological mutations, the novel creates a space for recognizing material agency. This aligns with Val Plumwood's advocacy for moving beyond Cartesian rationality to see "earth others as fellow agents and narrative subjects" within a "dialogical conception of self." This perspective is a step toward enhancing communication across species, fostering a more inclusive understanding of agency and interspecies relationships (Barca, 2024, p. 87).

Through its narrative, the novel underscores the intricate interplay between human and more-than-human agents, proposing that more-than-human elements actively influence socio-economic and political outcomes. This perspective challenges anthropocentric paradigms by suggesting that nature itself possesses agency. By illustrating the interdependence between human fate and the fate of species like the venomous lumpsucker, the narrative emphasizes the role of nature in shaping human destinies.

Latour frequently discusses the distribution of agency among various actors, emphasizing the necessity to "rematerialize our existence," which he describes as a process of "reterritorializing" or more aptly, "reterrestrializing" our connection to the Earth (Latour, 2017, p. 223). However, in placing Homo sapiens in the subject position, he seems to inadvertently replicate the subject-object relationship he strives to undermine. Instead of terrestrializing Homo sapiens, he seems to reiterate, or rather superimpose, the "bifurcation between the real—external, objective, and knowable—and the inside—unreal, subjective, and unknowable" (Latour, 2018). Agency emerges from the interaction of various agents and actants and is inherently distributed among them. People do not bestow agency but rather need to recognize and acknowledge how agency is dispersed across interconnected and animated material networks that include humans. In this sense, humans are not the subjects distributing agency but rather objects of this distribution, alongside other entities.

Beauman takes a cautious approach by acknowledging the inherent agency within networks, presenting a more integrated understanding where human and more-than-human entities coexist within a symbiotic web of influence and interaction. Rather than relying on representational logic to speak for more-than-human entities, Beauman allows their agency to emerge from their actions, demonstrating their impacts on both human and more-than-human life. The narrative does not depict humans as the primary agents distributing agency to more-than-human entities but rather as participants within networks of material agencies. It seeks to depict how individual lives, as well as human artifacts and institutions, are influenced by material agencies. Human beings remain far from the sole agents in this process; their agency is both enhanced and constrained by material agency. A similar theme is present in *Dr. Strangelove*, where the agency of the doomsday machine encroaches upon and dominates human agency, illustrating the complex interplay between human and material actants. Instead of trying "to give voice to the things that concern us" (Latour, 2017, p. 275) by drawing on representational logic, these narratives just show material agencies unfold in action through their consequences and the impacts they exert on human and more-than-human lives.

Ned Beauman's work underscores a philosophy of vibrant earthbound materiality, where material agencies actively shape both human and more-than-human events. This heralds a new materialist approach that maps interspecies relationships and extends these insights to human social, political, and economic constructs. By highlighting the impact of nonhuman species, environmental factors, and vibrant matter on human systems, Beauman suggests that the more-than-human can challenge and reshape anthropocentric constructs like capitalism. Humans, while responsible for exacerbating anthropogenic ecological mutations, are also integral to the resistance against these systems. Beauman even embraces radical actions to push back against the status quo, as evidenced by militant and violent reactions, including bio-bank disruptions, cloned tumors and gun attacks that destabilize the system's foundations. The narrative implies that the status quo will not change unless radically challenged by both human and material agencies to realize emancipatory worlding.

## References


Appadurai, A. (1986). Introduction: Commodities and the politics of value. In A. Appadurai (Ed.), *The social life of things: Commodities in cultural perspective* (pp. 3–63). Cambridge University Press. https://doi.org/10.1017/CBO9780511819582.003

Barca, S. (2020). *Forces of reproduction: Notes for a counter-hegemonic Anthropocene.* Cambridge University Press. https://doi.org/10.1017/9781108768380

Barca, S. (2024). *Workers of the earth: Labour, ecology and reproduction in the age of climate change.* Pluto Press.

Battistoni, A. (2025). *Free gifts: Capitalism and the politics of nature.* Princeton University Press.

Beauman, N. (2022). *Venomous lumpsucker.* Soho Press.

Besussi, A. (Ed.). (2017). *A companion to political philosophy: Methods, tools, topics.* Routledge.

Bevis, M. (2013). *Comedy: A very short introduction.* Oxford University Press.

Bonneuil, C., & Fressoz, J.-B. (2016). *The shock of the Anthropocene: The earth, history and us.* Verso Books.

Boykoff, M. (2019). *Creative (climate) communications: Productive pathways for science, policy, and society.* Cambridge University Press. https://doi.org/10.1017/9781108164047

Branch, M. P. (2014). Are you serious? A modest proposal for environmental humor. In G. Garrard (Ed.), *The Oxford handbook of ecocriticism* (pp. 378–391). Oxford University Press. https://doi.org/10.1093/oxfordhb/9780199742929.013.021

Caracciolo, M. (2022). *Contemporary fiction and climate uncertainty: Narrating unstable futures.* Bloomsbury Academic.

Caracciolo, M., Crockford, S., Lambert, S., & Ulstein, G. (2021). Phenomenology of everyday climate: An ethnographic approach to metaphor, affect, and the nonhuman. *ISLE Interdisciplinary Studies in Literature and Environment, 30*(2), 353–376.

Chandler, K. R. (2014). Poisonwood persuasion: Rhetorical roles of humor in environmental literature. *Interdisciplinary Literary Studies, 16*(2), 326–347. https://muse.jhu.edu/article/554074

DiCaglio, J. (2015). Ironic ecology. *Interdisciplinary Studies in Literature and Environment, 22*(3), 447–465. https://doi.org/10.1093/isle/isu106
Estlund, D. (Ed.). (2012). *The Oxford handbook of political philosophy*. Oxford University Press.
Foucault, M. (1975). *Discipline and punish: The birth of the prison*. [English edition published in 1977]. Pantheon Books.
Fraser, N. (2022). *Cannibal capitalism: How our system is devouring democracy, care, and the planet—and what we can do about it*. Verso.
Fraser, N. (2016). Expropriation and exploitation in racialized capitalism: A reply to Michael Dawson. *Critical Historical Studies, 3*(1), 163–178. https://doi.org/10.1086/685814
Fressoz, J.-B. (2024). *More, more, more: A history of the world's excesses*. Penguin Books.
Garrard, G. (2023). *Ecocriticism* (3rd ed.). Routledge.
Ghosh, A. (2024). *Smoke and ashes: Opium's hidden histories*. John Murray.
Grimbeek, M. (2017). *Margaret Atwood's environmentalism: Apocalypse and satire in the MaddAddam trilogy*. Karlstad University Studies.
Guo, X., Xu, B., Yu, H., Burnett, W. C., Li, S., Lian, E., Zhu, Z., Zhao, S., Chen, G., Duan, X., Dimova, N., Wang, Y., Zhuang, G., & Yu, Z. (2022). Exploration of deep ocean ferromanganese nodule fields using radon as a tracer. *Geophysical Research Letters, 49*(22), 100726. https://doi.org/10.1029/2022gl100726
Hamouchene, H. (2023). The energy transition in North Africa: Neocolonialism again! In H. Hamouchene & K. Sandwell (Eds.), *Dismantling green colonialism: Energy and climate justice in the Arab Region* (1st ed., pp. 29–48). Pluto Press. https://doi.org/10.2307/jj.7583922.7
Hardin, G. (1968). The tragedy of the commons. *Science, 162*(3859), 1243–1248. https://doi.org/10.1126/science.162.3859.1243
Heise, U. K. (2016). *Imagining extinction: The cultural meanings of endangered species*. The University of Chicago Press.
Horkheimer, M., & Adorno, T. W. (2016). *Dialectic of enlightenment*. Verso.
James, E. (2024). Fiction, belief, and climate change: Paratexts, skeptics, and objects of care. In *The Routledge handbook of fiction and belief* (pp. 323–335). Routledge. https://doi.org/10.4324/9781003119456-28
Latour, B. (1999). *Pandora's hope: Essays on the reality of science studies*. Harvard University Press.
Latour, B. (2017). *Facing Gaia: Eight lectures on the new climatic regime*. Polity.
Latour, B. (2018). *Down to Earth: Politics in the new climatic regime*. Polity.
Malafouris, L. (2013). *How things shape the mind: A theory of material engagement*. MIT Press.
Malm, A., & Zetkin Collective. (2021). *White skin, black fuel: On the danger of fossil fascism*. Verso.
Meeker, J. W. (1974). *The comedy of survival: Studies in literary ecology*. Scribner.

Morris, C. W. (Ed.). (2003). *The social contract theorists: Critical essays on Hobbes, Locke, and Rousseau*. Rowman & Littlefield.

Peiffer, K. S. (2000). *Coyote at large: Humor in American nature writing*. The University of Utah Press.

Riofrancos, T. (2025). *Extraction: The Frontiers of Green Capitalism*. W. W. Norton.

Robinson, C. J. (2019). *Cedric J. Robinson: On racial capitalism, black internationalism, and cultures of resistance*. Pluto Press. https://doi.org/10.2307/j.ctvr0qs8p

Rupasinghe, R., Chomel, B. B., & Martínez-López, B. (2021). Climate change and zoonoses: A review of the current status, knowledge gaps, and future trends. *Acta Tropica, 226*, 106225. https://doi.org/10.1016/j.actatropica.2021.106225

Rushdie, S. (1988). *The satanic verses*. Vintage Books.

Seymour, N. (2018). *Bad environmentalism: Irony and irreverence in the ecological age*. University of Minnesota Press.

Sørensen, M. (2016). *Humor in political activism creative nonviolent resistance*. Palgrave Macmillan.

Traub, C. (2018). From the grotesque to nuclear-age precedents: The modes and meanings of cli-fi humor. *Studies in the Novel, 50*(1), 86–107. https://www.jstor.org/stable/48559258

Varoufakis, Y. (2023). *Technofeudalism: What killed capitalism*. Penguin.

Wagner, P. (2023). Embracing the environmental grotesque and transforming the climate crisis. *ISLE: Interdisciplinary Studies in Literature and Environment, 30*(4), 911–930. https://doi.org/10.1093/isle/isab086

Warren, K. J. (2000). *Ecofeminist philosophy: A Western perspective on what it is and why it matters*. Rowman & Littlefield.

Wiek von Mossner, A. (2020). Affect, emotion, and ecocriticism. *Ecozon@ European Journal of Literature Culture and Environment, 11*(2), 128–136.

Zekavat, M. (2019). Reflexive humour and satire: A critical review. *European Journal of Humour Research, 7*(4), 125–136. https://doi.org/10.7592/EJHR2019.7.4.zekavat

Zekavat, M. (2025). Communicating multispecies planetary health through multimodal humour and satire. *Medical Humanities*, medhum-013359. https://doi.org/10.1136/medhum-2025-013359

Zekavat, M. (Forthcoming-b). Planetary health and emancipatory worlding. *Global Health Promotion*. http://dx.doi.org/10.1177/17579759251378432

CHAPTER 7

# Emancipatory Worlding

In a travel agency adorned with images of iconic tourist attractions such as the Eiffel Tower and the Brandenburg Gate, a woman walked in with an air of determination. She approached the man standing behind the counter, her gaze unwavering. With a wry smile, she asked, "*which destinations are also shifting to fascism, so I can respond to their jabs by laughing at their hypocrisy?*" (Dai, 2024).

In this book, I attempted to argue that satire can serve critical environmental advocacy. While it offers a unique avenue for engaging environmental issues, questions remain about its efficacy and scope. Can the act of "responding to jabs by laughing at hypocrisy" confront denialism, skepticism, vested interest, and truly drive meaningful change? The analysis suggests that while satire alone may not suffice, it holds substantial power when thoughtfully applied.

Both apocalyptic narratives and satirical productions are susceptible to commodification under the culture industry, yet they serve distinct purposes and provoke different responses. Commodified apocalyptic narratives typically capitalize on fear, with a limited capacity for self-reflection, as a self-reflexive approach to fear-inducing narratives risks dissipating the suspense and fear and losing its impact. Apocalyptic narratives, therefore, can hardly foreground the process that goes into their emotional manipulation and monetization.



M. Zekavat, *Leveraging Satire for Environmental Advocacy*,
https://doi.org/10.1007/978-3-032-04838-7_7

Satire diverges from apocalyptic narratives in three ways. First, it is easier to imagine and find instances of satire that refuse commodification and monetization even under capitalism: One can think of graffiti art; latrinalia; grassroots protest satire as in hand-painted banners, puppets, or costumes at climate marches or anti-war rallies; zines and samizdat as in underground political satire in pamphlets, or hand-distributed comics such as feminist or queer zines of the 1980s–1990s, or Soviet-era samizdat that used biting satire to mock the regime; street theatre and guerrilla performance like fake corporate presentations; vernacular digital satire as memes, particularly those that avoid viral platforms or delete themselves after a short lifespan; and indigenous and local oral satire performed in rituals or gatherings where the satire is context-bound and not recorded or sold.

Second, in contrast to apocalyptic narratives, commodified satire thrives on introspection. Not only does it not lose its thrill, but it can even turn more playful and engaging by foregrounding and undermining its own positionality. Moreover, satire can cleverly mock its own status as a monetized commodity, wielding subsequent awareness to enhance both its critique and appeal.

Third, satire doubles down on its dual nature instead of concealing its complicitous critique. It locates itself both inside and outside the dominant discourse of society and does not hesitate to come clean about being both contestatory and complicit. It is aware of its ambivalent socioeconomic standing as "being totally complicitous with capitalism, while (implicitly) positioning itself safely outside it" (Hutcheon, 1988, p. 212). While the culture industry commodifies satire, turning it into a means of social control, and capitalism manipulates emotions and desires through satire to maintain subjugation, satire can occasionally evade commodification, counter passionate exploitation and servitude under capitalism, or provide an act of scornful defiance in its complicitous critique (Cf. Lordon, 2014).

Throughout the book, I detailed how satire can transcend traditional anthropocentric narratives and the logic of representation. By acknowledging and critiquing the interplay of racial, gender, class, and colonial privileges across various temporal and spatial contexts, satire can serve as a transformative force in environmental discourse. Moreover, by denouncing imperialist frameworks and embracing decolonial and materialist perspectives, satire can contribute to a more inclusive and equitable environmental advocacy.

The effective use of satire in environmental advocacy necessitates a comprehensive understanding of several critical elements. Attention must be paid to the creator, recipients, and their interactions. Understanding the recipient's perspectives, predispositions, and status in power relations allows for tailoring the message to engage the audience effectively and provoke thought. The content itself should be meticulously crafted to balance satire with poignant critique, while remaining sensitive to the complexities and nuances of the issues at hand. The successful communication of the message might require the use of a communicator or persona that is different from the satirist creator. Besides the skills required to convey the satirical message to resonate with the target audience, the communicator could be selected to regulate their emotional and cognitive responses. A conservative persona might be more effective in reaching out to conservative audiences or ironically undermining conservative arguments.

Moreover, the materiality of communication, like choosing an appropriate medium and distribution outlet, is vital for maximizing the reach and impact of the message. Visual and multimodal satire might be more effective in reaching certain audiences as are outlets like different forms of traditional media or digital social media. The context in which satire is delivered can significantly influence its reception; thus, it is essential to consider cultural, social, and political factors that may affect the audience's interpretation. By meticulously addressing these components, satire can be wielded as a powerful tool in the pursuit of meaningful environmental advocacy. These can ensure satire to be an invaluable component of a multi-faceted approach to environmental activism, capable of provoking thought, challenging norms, and inspiring action across diverse audiences. Despite all these, satire falls short of being a panacea.

Besides promoting environmental satire, I advocated for a decolonization of hegemonic psychology in understanding the impacts of satire on environmental behavior, whether at the individual, corporate, or collective level. This approach emphasizes the need to focus on the roles of corporations, politicians, and systemic issues, rather than disproportionately attributing blame to individuals. We need to recognize that diverse communities contribute to anthropogenic ecological mutations in varying degrees and possess unequal resources to address them. We also need to be aware that the burdens and benefits of 'green transitions' are not equitably distributed.

Furthermore, I critiqued the corporate dynamics inherent within the neoliberal teaching machines of academia, urging these institutions to reflect on their ideological functions, allegiances, and connections to specific interests in ecological crises that inevitably intersect with other sociopolitical concerns whether it be the way they treat students and faculty members who raise their voice against genocide and omnicide, their investment strategies, or their financial sources.

These can be meaningful steps toward promoting environmental and multispecies justice. However, the question remains whether aspiring for multispecies justice is the most effective strategy for navigating the complexities of the brave new world of techno-feudalism.

## Inexorable Miscarriages of Justice

In the bustling corridors of a Colorado courthouse, an unusual case unfolded. A group of animal rights activists had filed a lawsuit on behalf of Missy, Kimba, Lucky, LouLou, and Jambo, elephants residing in the local zoo. Invoking habeas corpus—a legal process allowing individuals in custody to challenge their detention—the activists sought to have the elderly elephants relocated to a sanctuary.

The Colorado Supreme Court, however, ruled unanimously with a 6–0 decision, focusing on the legal question of whether an elephant could be considered a person. State Supreme Court Justice Maria Berkenkotter, in her ruling, noted, "it bears noting that the narrow legal question before this court does not turn on our regard for these majestic animals generally or these five elephants specifically; because an elephant is not a person, the elephants here do not have standing to bring a habeas corpus claim" (Pannett, 2025).

On January 31, 2025, *The Onion* published its satirical take on the story in the form of three responses from:

> "Molly Klink, Jam Spreader" who asks, "We can still sue elephants though, right?"
>
> "Joe Caruso, Megaphone Tester" who maintains, "That's all the greenlight I needed to start eating them."
>
> And "Will Szymanski, Water Bottler" who wonders, "What's more human than suing people?"

Satire, deeply connected to the ancient Greek concept of parrhesia (παρρησία), embodies the act of speaking truth to power. It is often lauded for its capacity to "punch up" rather than "punch down," challenging authority and renegotiating or even upending power structures. This characteristic has led many satirists to face severe repercussions, including imprisonment or forced exile by authoritarian regimes. Such examples underscore satire's potency as a tool for resistance and its role in the broader struggle for justice. As Chiew et al. (2024) observe, humor is an underappreciated yet vital element in organizing movements toward a more equitable world; however, they caution that while satire and humor can contribute to pleasurable and just outcomes, these effects are not guaranteed. Nonetheless, the courage and strength demonstrated by satirists highlight the enduring power of satire in advocating social change and challenging oppressive structures.

Besides advancing environmental justice, satire possesses the potential to further multispecies justice due to its inherent corporeal, scatological, and material entanglements. These elements allow satire to address complex interdependencies between humans and other species in a manner that is both critical and engaging. Heise (2016), in her celebrated work *Imagining Extinction: The Cultural Meanings of Endangered Species*, notes that "only occasionally has conservationist writing mobilized the resources of comedy, a genre that offers an alternative model for thinking about biodiversity. Comedy emphasizes contingency and improbable modes of survival over predictability and extinction and thereby opens up different cognitive and emotional attachments to the lives of other humans as well as nonhuman species" (Heise, 2016, p. 14). By embracing the unexpected and fostering a sense of interconnectedness with all forms of life, satire can cultivate a more inclusive and empathetic understanding of biodiversity. This approach encourages audiences to reconsider their relationships with the natural world and supports the pursuit of justice for all species.

Heise also argues for cosmopolitanism as a more suitable framework than universalism. She references the scholarly discourse that has emerged around Dipesh Chakrabarty's work on climate change, which underscores the challenges of forming a unified human perspective that includes nonhuman species. This discourse emphasizes the necessity of a cosmopolitan approach, one that carefully and deliberately redefines human identity through diverse cultural, technological, and political prisms. Speculative fiction, Hiese argues, illustrates this by transitioning from narratives of human dominance and species extermination to visions of cooperative multispecies societies. These fictional scenarios expand the imaginative and narrative possibilities for cosmopolitanism, advocating for multispecies justice by encouraging audiences to imagine futures where humans and other species collaborate through shared agreements and communities.

In discussing the intricacies of multispecies justice within the context of conservation, Heise notes that "environmentalists currently face the challenge of reenvisioning conservation in terms that enable the imagination not so much of the end of species as of their future" (Heise, 2016, p. 50). Satire plays a crucial role in this reimagining by leveraging its emotional impact to question human exceptionalism and foster interspecies solidarity. The strength of satire lies in its departure from predictable doom-and-gloom narratives, which have lost much of their rhetorical power due to their repetitiveness. Instead of depicting animals merely as endangered treasures of beauty or biological information, satire presents them as ordinary beings navigating life's complexities, often unsuccessfully, much like humans. By drawing parallels between animal behaviors and human technologies, satire challenges the notion of human exceptionality, even when differences are evident. These comparisons highlight the shared struggles between human and nonhuman protagonists, inviting empathy and a sense of camaraderie. Ultimately, satire suggests that by recognizing our mutual challenges in an often inhospitable world, we can build solidarity and ease the collective struggle for survival (Heise, 2016).

Satire can further offer a fresh perspective on the relationship between modernization and nature, distinct from the elegiac or tragic narratives often used to address species endangerment and extinction. As Heise (2016) notes, a comedic approach frames nature and culture as parallel and intersecting histories of experiments that both succeed and fail. This perspective does not depict nature as a pristine realm disrupted by modernity, but rather as an evolving landscape where extinction signifies a failure to be prevented, yet also a catalyst for new beginnings. The urgency for conservation is thus grounded in a narrative that values biological diversity and acknowledges shared vulnerabilities. Such an approach redefines extinction not as an endpoint, but as an opportunity for new futures and evolving life forms.

Heise therefore calls for a shift away from apocalyptic tropes, highlighting that this transition might introduce equally somber themes of endangerment. She suggests that environmentalists, in moving from narratives of apocalypse and the end of nature, might adopt tropes of "oppression, slavery, and holocaust" (Heise, 2016, p. 160).

The question remains if justice is enough to bring about such a shift to these tropes. I argue that emancipation might be a more deserving pursuit. Justice, often perceived as passive, may not suffice to catalyze the necessary shift toward addressing themes of oppression, slavery, and (slow) violence. Instead of passive recipients, these projects require "active agents of refusal, resistance, and world-making" (Bosch, 2025, p. 1715). In contrast, emancipation paves the way for revolutionary action, offering a more urgent and dynamic response for the marginalized and exhausted of the earth. Unlike justice, which is often polemic and deliberative, emancipation is inherently political and can be militant, driven by the need for immediate and drastic change. It does not tolerate the unfolding slow violence that often accompanies systemic injustice but instead seeks to halt it without delay. It is acutely aware that epistemic subjugation and bondage exacerbate material disenfranchisement and dispossession. Emancipation demands immediate restitution and challenges even the provisional acceptance of suffering and delay,

> For who would bear the whips and scorns of time,
>
> Th' oppressor's wrong, the proud man's contumely,
>
> ..., the law's delay,
>
> The insolence of office, and the spurns
>
> That patient merit of th' unworthy takes ... ? (William Shakespeare's *Hamlet*, III.I.71–75)

Justice is often pursued through traditional channels and institutions, which have historically prioritized the interests of a select few over the majority. The institutions designed to uphold justice frequently perpetuate systemic injustices, as against people of color and marginalized ethnicities. They are susceptible to corruption, political, and financial influences. The system is undermined by indulgences such as luxurious trips and expensive RVs, acts of pardon or undue pressures, leading to a sluggish response that fails to maintain even the illusion of justice they purport to uphold. This is evidenced by instances where significant legal actions, like those against political figures such as Donald J. Trump, are quietly dismissed or overlooked following his electoral victories. These institutions remain vulnerable to the sway of lobbyists and political considerations, often riddled with immunities, exceptions, and loopholes that can be exploited, undermining their credibility and effectiveness in delivering justice. These are just some pitfalls of justice in 'democratic' governments, let alone many authoritarian regimes where the despot is the law. These flaws are flagrant enough to invoke medical professionals to observe that "the law may be an ass…" (Abbasi, 2025).

Justice, when pursued through traditional and institutional channels, often reflects and perpetuates the biases embedded within these structures. The discourse on environmental justice, for instance, is frequently criticized for being Western and Eurocentric, with deeply ingrained imperialistic and colonial biases. This raises significant doubts about the entelechy of justice. An illustrative example is the commitment made during COP29, where wealthy nations pledged US$300 billion by 2035 to assist poorer countries in addressing climate change. While this decision, made at the COP in Azerbaijan, aligns with the rhetoric of justice, it remains uncertain whether these insufficient pledges will be ever fulfilled. Although it represents a step toward justice, it falls short of achieving liberation. Instead, it risks replicating existing power dynamics under a new guise, perpetuating eco-colonialism rather than dismantling it.

Carbon offset markets exemplify the limitations and pitfalls of current justice frameworks, revealing that they are far from zero-sum solutions. Many cherish the misconception that they can negate their carbon footprint through minimal financial contributions, such as offsetting the emissions of extensive long-haul flights for a nominal fee. This belief extends to assumptions that poor nations benefit from this "easy money," that nature thrives on these investments, and that participants become environmental heroes. This perception is inherently misleading, though. In reality, carbon markets often facilitate land grabs reminiscent of colonial practices, albeit cloaked in a superficial green narrative. These markets lead to the displacement and dispossession of local communities, bolster authoritarian regimes, and disrupt ecological balance. By externalizing environmental damages from wealthy nations to poorer regions, they fail to liberate but instead further subjugate these areas to a Western-centric discourse of justice, perpetuating rather than resolving systemic inequities (see Sultana, 2024).

Likewise, politicians frequently tout "just and inclusive transitions," yet this phrase often masks (might be designed to mask?) underlying miscarriages of justice. While the Just Transition strategy appeared promising in showcasing labor's ability to plan for a post-carbon economy, it largely excluded unwaged and reproductive labor—i.e., women constituting more than half of the global working class—thereby significantly constraining its political potential. Trade unions and workers are forging a new path in the history of labor environmentalism, one in which the concepts of green growth and Just Transition offer economic prosperity and stability akin to the aspirations of the Fordist era. However, embracing this vision will not shield organized labor from the structural weaknesses and limitations that have severely diminished its power in many nations. On the contrary, by endorsing capital's "green" restructuring of the global economy, trade unions risk positioning themselves in opposition to peasant and Indigenous communities, landless rural laborers, unpaid domestic and care workers, subsistence farmers, and others who bear the burdens of "green" capitalism—perpetuating cycles of dispossession and subjugation (Barca, 2024, p. 16).

Moreover, justice systems are predominantly anthropocentric, focusing primarily on human interests and rights. While there are notable exceptions, such as the enshrined rights of Pachamama in the Bolivian constitution, most juridical frameworks do not adequately recognize, engage with, or protect the rights of non-human entities or the material environment. When legal cases do address natural resources or environmental issues, they often do so from the perspective of their utility to humans, rather than acknowledging their intrinsic value. This anthropocentric bias is deeply entrenched, limiting the scope of justice to human-centered concerns and failing to extend protection or rights to the broader ecological community as in the case of the Colorado Supreme Court's decision.

Thus, justice is an ideal, not in the sense that we merely need to strive harder to achieve it, but rather in the sense that it remains inherently unattainable. It parallels Lacan's concept of the Real, representing an ultimate lack—unknowable and resistant to full articulation or realization. Like the elusive pursuit of happiness described in Samuel Johnson's *The History of Rasselas, Prince of Abissinia*, where "each believes it possessed by others, to keep alive the hope of obtaining it for himself," justice serves as a concept perpetually out of reach. It intrudes upon our consciousness through its evasive returns and disruptions, reminding us of its presence without ever fully materializing.

Justice, as it is currently conceived and practiced, often obscures and conceals more than it divulges. Its short-lived actions have repeatedly proven inadequate in addressing the enduring consequences of systemic injustices. As Bob Dylan aptly puts it, "justice is a game," one that the marginalized and the Exhausted of the Earth increasingly refuse to play. This "game" has effectively relegated regions like Palestine and Western Sahara to a state of erasure, highlighting the limitations and failures of the justice project. The persistent inability of justice systems to address and rectify these long-standing issues suggests that the project of justice, as it stands, is fundamentally flawed and inadequate in achieving genuine and lasting restitution. It is a project that is always already failed.[1]

This refusal to engage with the failed and failing game of justice opens up a space for emancipation. As Derrida interprets Kafka's "Before the Law," the law derives its legitimacy from individuals appealing to it, even as it withholds justice. By stepping away from this unproductive cycle, individuals can defy the structures that perpetuate injustice and create new opportunities for emancipation.

In his deconstruction of Franz Kafka's parable, Derrida explores the notion that the law is not a fixed or accessible entity but rather something perpetually deferred. This understanding is informed by Derrida's broader theory of *différance*, which posits that meaning is never fully present but continuously postponed through an endless interplay of signifiers. In Kafka's narrative, a man seeks entry into "the law," only to be perpetually denied access by a gatekeeper who suggests that entry might be possible later, but not now. This is exactly how justice operates: it is never fully accessible or graspable, existing instead in a constant state of deferral.

While Horkheimer and Adorno (2016) critique how "justice is subsumed in law" after the Enlightenment, Derrida strives to distinguish between law and justice, viewing the former as an institutional structure with rules, while justice remains an infinite, unknowable, and unattainable ideal. Because justice cannot be codified or fully realized, the law is inherently incomplete, striving for order without ever achieving true justice.

The failures of justice have not escaped the scrutiny of satirists; indeed, satire has long served as a potent medium for exposing and condemning such miscarriages. A striking example emerged on September 8, 2025, when the renowned street artist Banksy unveiled a provocative mural on

[1] Cf. Rohan Chakrabarty's cartoon on "ICJ Ruling on Climate Change" at https://greenhumour.substack.com/p/icj-ruling-on-climate-change

the exterior wall of the Queen's Building at the Royal Courts of Justice in London. The artwork, shared on Banksy's Instagram with the caption "Royal Courts of Justice. London," depicts a judge—clad in the traditional wig and black robes of the English judiciary—striking a defenseless protester with a gavel. The protester, lying prone on the ground, holds a placard splattered with red blood, the only splash of color in an otherwise monochrome composition. The mural was widely interpreted as a searing critique of the legal suppression of Palestine Action, a protest group targeted under the Terrorism Act 2000—a law that, in its application, echoes the colonial logic of criminalizing resistance to oppression. Its placement on the Royal Courts of Justice was deliberate, not only amplifying its symbolic weight but also implicating the British state in a legacy of imperialist violence. The judiciary, historically complicit in upholding the structures of colonialism, here becomes a visual metaphor for the ongoing erasure of Palestinian rights—a continuum from the Nakba to the present. Yet, the artwork was swiftly covered and later removed, a move that only deepened the irony of its message: the very institutions that claim to uphold justice are the ones silencing its critique.

The layers of irony are unmistakable. First, the image portrays a judge, the very embodiment of justice, inflicting violence upon an unarmed individual exercising their right to protest against genocide. Second, the scene subverts the ideal of justice, replacing it with brutality. Most damning of all, the mural's erasure from the walls of the Royal Courts of Justice underscores the institution's inability—or refusal—to confront its own complicity in injustice. In this act of censorship, justice itself becomes the perpetrator, silencing the critique it cannot endure.

Another incisive example of satire exposing systemic multispecies injustice appeared on *The Late Show with Stephen Colbert* on September 26, 2025. During the show's "Meanwhile" segment, Colbert reported on a groundbreaking medical achievement: a human patient had survived for six months after receiving a kidney transplant from a pig, as documented in the prestigious journal *Nature*. The announcement was framed as a triumph of scientific progress, a narrative of human ingenuity and survival.

Yet Colbert swiftly subverted this celebratory tone with a biting question: "That's fantastic. How's the pig doing? Is he [*sic*] okay? Because that was really nice of that pig. I'd like to send him something. ... I'm sure it was just one kidney; they let the pig keep the other kidney ... or they put the pig on dialysis. I mean, they do something nice for the pig. They're not monsters." In an instant, the focus shifted from human achievement to the unseen cost borne by the pig—a sentient being reduced to a mere resource. What had been presented as a medical milestone suddenly revealed itself as an act of extraction, where one life is extended at the expense of another's suffering. Colbert's satire lays bare the ethical blind spots of scientific progress, forcing the audience to confront the asymmetrical power dynamics at play. The pig, stripped of agency and dignity, becomes a silent victim in a narrative that celebrates human advancement while ignoring the moral weight of multispecies exploitation. The joke, then, is not just humorous—it is a provocation, exposing the uncomfortable truth that progress often comes at a hidden, and deeply unjust, cost.

In proposing emancipatory worlding as an alternative to justice, I draw on the insights of thinkers like Marx, Spivak, and Ferdinand. I approach the term "decolonization" with caution, recognizing that its frequent overuse has led to co-option and a dilution of its original meaning. I am mindful of instances where a small group of WEIRD (Western, Educated, Industrialized, Rich, and Democratic) individuals engage in cosmetic activities, such as a brief yoga session, and label them as efforts to "decolonize" health. These actions risk trivializing the profound and complex work required for genuine decolonization and liberation. As Nkrumah (1965, p. 34; Cf. Frank, Hironaka, & Schofer, 2000) clarifies,

> Decolonization is a word much and unctuously used by imperialist spokesmen to describe the transfer of political control from colonialist to African sovereignty. The motive spring of colonialism, however, still controls the sovereignty. The young countries are still the providers of raw materials, the old of manufactured goods. The change in the economic relationship between the new sovereign states and the erstwhile masters is only one of form. Colonialism has achieved a new guise. It has become neo-colonialism, the last stage of imperialism; its final bid for existence, as monopoly capitalism or imperialism is the last stage of capitalism. And neo-colonialism is fast entrenching itself within the body of Africa today through the consortia and monopoly combinations that are the carpet-baggers of the African revolt against colonialism and the urge for continental unity.

I agree with Bosch (2025, p. 1715–1716) who insists that "to decolonize is not to diversify the canon or retrofit existing paradigms with more inclusive language. It is to rupture the architectures of domination that structure our field, intellectually, institutionally and economically. … It is to materially disrupt the infrastructures—epistemic, technological, institutional—that reproduce empire." Above all, one should not forget that "as

Marx taught in *On the Jewish Question*, decolonization is not human emancipation; nor does it necessarily lead in the direction of human emancipation" (Burawoy, 2025, p. 256). Emancipatory worlding, therefore, seeks to go beyond empty gestures, aiming for systemic change and a reimagining of interspecies and human social relations.

Emancipatory worlding, as I conceive it, is informed by Malcom Ferdinand's notion of decolonial ecology. This approach challenges the violence imposed on both humans and more-than-humans through colonial practices of inhabitation. Decolonial ecology critiques environmental colonialism, where certain states or groups directly or indirectly manipulate Earth's resources for profit, leading to environmental degradation for local inhabitants. It opposes a mode of living that exploits planetary resources for the enrichment of a privileged few, while leaving entire populations in states of food insecurity and environmental degradation.

Decolonial ecology also addresses the heterotopic legacy of colonization, where certain peripheral spaces are treated differently from the metropole in being assigned an inferior status or misrepresented as empty land, terra nova, justifying the rise of sacrifice zones, outsourcing of ecological disasters, displacing suffering and violence. This violence is of course tacitly endorsed by governments, corporations, and consumers who knowingly or unknowingly perpetuate and reinforce it by indirectly exploiting indentured labor and remote sacrifice zones as long as they can afford to continue consuming and while their immediate environment is safe.

Decolonial ecology exposes the socio-political structures that alienate people from their bodies, the Earth, and the world. It is an ecology of struggle, challenging colonial methods of inhabiting the Earth and advocating for new ways of living together. Confronting ecosystem destruction is thus intertwined with the pursuit of emancipation, representing a dual healing process that reimagines both decolonization and resistance to environmental degradation.

Emancipatory worlding emphasizes the interconnectedness of all forms of oppression, including those based on gender, sexual identity, race, ethnicity, religion, and class. It stands in opposition to imperialism, (settler) colonialism, apartheid, militarization, and all forms of capitalism. This approach refuses to separate colonial relations and reparations from environmental degradation, anthropogenic ecological mutations, and multispecies agency. It encompasses "political, epistemic, scientific, legal, and philosophical struggles to dismantle the colonial structures of living together and the ways of inhabiting the Earth that still maintain the domination of racialized [and colonized] people, particularly women" (Ferdinand, 2022, p. 14), through financial, military, or other forms of coercion.

It recognizes that the enduring impacts of history cannot simply be wiped away through legislation or by accelerating technological advancement. As Ghosh notes, "the stamp of the past sometimes sinks so deep into the fabric of everyday life that its traces are difficult, if not impossible, to erase" (Ghosh, 2024, Ch 7). Emancipatory worlding deliberately embraces politics and strategies that acknowledge and engage interconnected histories.

Emancipatory worlding fundamentally departs from the colonial principles of inhabitation, which are characterized by geographical subordination, exploitation, extraction, expropriation, land grabs, massacres, ethnic cleansing, land clearing, and "othercide"—the denial of coexistence with those who are different from oneself. It is based on the understanding that "prohibiting the domination and exploitation of human beings by other human beings through slavery, the slave trade, and colonization is illusory while a social and economic organization is maintained whose function is the colonial exploitation of the Earth. Changing politics entails changing ecology. … [Moreover] protecting the Earth's natural spaces and forests from the financial desires of certain human beings is illusory as long as we accept the servitude of other human beings through enslaving and colonial domination. Changing ecology entails changing society" (Ferdinand, 2022, pp. 123–124). It underscores the inseparable links between ecological and social transformations, human and more-than-human, to achieve liberation.

Emancipatory worlding firmly rejects strategies and schemes that transform semi-natural disasters into profitable opportunities, thereby reinforcing the world's colonial, capitalist, and consumerist structures, enriching those in power, and worsening the plight and subjugation of marginalized groups. Emancipatory worlding opposes the commodification and commercialization of every aspect of life. Instead of relying on geoengineering, bioengineering, or superficial measures like reforestation, rewilding, or the reintroduction of species, which often remains disconnected from socio-economic and political realities, emancipatory worlding calls for a more integrative approach. It emphasizes holistic and systemic change that addresses the root causes of environmental and social issues, moving beyond cosmetic solutions.

In its approach, emancipatory worlding draws on systems of belief, sciences, and technology but transcends them all. It maintains a critical stance towards science, technology, and data, recognizing the existence of multiple regimes of truth and alternative epistemologies across both milieus and cultures. Emancipatory worlding respects ancient wisdom and employs modern science while acknowledging that Western science, despite its dominance, is neither inherently correct nor neutral. It is acutely aware that science, despite its veneer of objectivity, often serves specific interests. In the realm of discursive contentions, emancipatory worlding strives to create space for Indigenous and Aboriginal knowledges, treating them not as objects of observation or subjects of research but as valuable sources of insight and understanding.

Emancipatory worlding remains firmly grounded in the bodily, material, and the rootedness of human existence. It acknowledges this rootedness as “an opening, through the political action of a living together, of the infinite horizon of encounters and sharing” (Ferdinand, 2022, p. 17). It embraces and cherishes material agency and enmeshment, offering a new ontology that blurs the distinctions between human and non-human animals, vibrant matter, as well as technology. Emancipatory worlding embraces “Worldly-ecology,” which “assumes a relational ontology that recognizes that our existence and our bodies are made up of encounters with a plurality of human beings and a plurality of non-human beings. A creolized gestalt ontology or a gestalt creolization of the human species” (Ferdinand, 2022, p. 231).

Emancipatory worlding exhibits a mature appreciation of material entanglements and agency. It recognizes that nature is not inherently a victim, whether considering immobile resources like forests, mobile resources such as water, or fugitive resources such as wildlife. As Nixon (2011, Introduction) observes, “resistance may assume not just human forms but also arise from an unanticipated recalcitrance on the part of a targeted resource, which may prove harder to commodify and profitably remove or manage than corporate moguls foresaw.” This perspective is exemplified by the historical resistance of water to various privatization efforts. Emancipatory worlding acknowledges and celebrates these dynamics, understanding that the material world itself can be an active participant in resisting commodification and exploitation, thus challenging anthropocentric narratives and practices.

Satire could be a technique to embrace such vital materiality as it helps admit, what Adorno calls, “a ‘playful element “into one’s thinking and to be willing to play the fool. The negative dialectician" knows how far he remains from "knowing nonidentity," and yet he must always talk as if he had it entirely. This brings him to the point of clowning. He must not deny his clownish traits, least of all since they alone can give him hope for what is denied him'" (Bennett, 2010, p. 15). Most works examined in this book fall short of wholeheartedly embracing "the courage of a clown" (Bennett, 2010, p. 15) and a radical perspective.

There is a pressing need for further development and exploration of these ideas. Without such efforts, how can Palestinians—or any marginalized group—realistically hope for emancipation? Perhaps Edward Said intuitively recognized the limitations of traditional avenues for justice when he set aside his piano keyboard to throw stones in the occupied territories. The challenges of settler colonialism, ethnic cleansing, genocide, apartheid, slow violence, corporeal suffering, material scarcity, pollution, racialization, and dehumanization demand more than conventional responses. They require materialist tenacity to dismantle these entrenched systems of oppression.

It is not just cultural productions or social structures that fall short of the ideals of emancipatory worlding; policies often perpetuate and reinforce these shortcomings, sometimes even more alarmingly. Even agendas that claim to be progressive frequently fail to align with the most basic principles of emancipatory worlding. The European Green Deal serves as an example. It appears to be grounded in ecomodernist illusions, relying on finance capital and technology to address problems that these very elements have helped create in the first place. Market solutions, such as carbon pricing, are valorized, with the private sector, rather than commoning, positioned as key to financing the green transition. This approach risks reinforcing solutions that have previously backfired. Above all, the Deal is anthropocentric, parochial, and imperial in its ethos, failing to engage with the broader, interconnected challenges that emancipatory worlding seeks to address.

The European Green Deal appears to be explicitly anthropocentric and Eurocentric, as evidenced by its stated goals to "protect, conserve and enhance the EU's natural capital, and protect the health and well-being of citizens from environment-related risks and impacts. … It must put people first, and pay attention to the regions, industries and workers who will face the greatest challenges." This emphasis assumes it is both possible and desirable to prioritize human interests and well-being, potentially at the expense of a more integrated and holistic approach that considers the intrinsic value of all ecological components. By doing so, it denies material agency and fails to address the interconnectedness of human and more-than-human entities in pursuing emancipatory worlding.

While the Deal gestures toward prioritizing human well-being, it even falls short of genuinely fulfilling this claim. Instead, it places markets at the forefront, assuming that investment will naturally yield positive outcomes. This is evident in the section titled "Pursuing green finance and investment and ensuring a just transition," where at least 30% of the budget is allocated to "fighting climate change." However, the focus remains on finding ways to maintain the status quo through various strategies rather than pursuing a radical overhaul. This approach suggests a preference for incremental change within existing structures, rather than embracing the transformative changes necessary for achieving genuine and timely emancipatory worlding.

One example of maintaining the status quo is the emphasis on carbon pricing and other technological fixes. The Deal calls for "investment to provide affordable solutions to those affected by carbon pricing policies, for example through public transport, as well as measures to address energy poverty and promote re-skilling. Coherence of climate and environment policies and a holistic approach are often a precondition for ensuring they are perceived as fair, as illustrated by the debate on taxation of various modes of transport." While it stresses the importance of social coherence, it falls short of encouraging a fundamental restructuring of labor relations or advocating for universal basic income. Instead, it focuses on retraining labor to fit within existing structures, keeping workers in subordinate roles.

While it states that "at national level, the European Green Deal will create the context for broad-based tax reforms, removing subsidies for fossil fuels, shifting the tax burden from labour to pollution, and taking into account social considerations," it does so with the underlying belief that "well-designed tax reforms can boost economic growth and resilience to climate shocks and help contribute to a fairer society and to a just transition." This reliance on capitalist market mechanisms is not an attempt to redistribute wealth, promote multispecies commoning, or make reparations. Instead, it focuses on adjusting existing economic structures to maintain growth and foster resilience within a capitalist framework.

In its efforts to uphold existing socio-economic structures, the Deal also perpetuates European parochialism and imperialism. The historical context of colonialism and the EU's imperial projects, including the legacy of the slave trade and what is aptly termed the Plantationocene, is conspicuously absent from the discourse. There is no mention of restitution or system change; instead, the focus remains on finding new ways to maintain the status quo in Europe. The Just Transition Mechanism, for instance, aims to support regions and sectors most affected by the transition due to their reliance on fossil fuels or carbon-intensive processes. But it lacks a sincere engagement with the historical and systemic inequities that have contributed to the current environmental and social urgancies.

In the Deal, the non-EU "other" is referenced only fleetingly, with the focus shifting from a parochial perspective to a global one primarily in the context of EU security concerns. The document acknowledges that "global climate and environmental challenges are a significant threat multiplier and a source of instability," suggesting that the ecological transition will impact geopolitics, including European economic, trade, and security interests.

While the EU commits to collaborating with global partners to enhance climate and environmental resilience, aiming to prevent these challenges from escalating into conflicts, food insecurity, population displacement, and forced migration, the sincerity of this commitment to "support a just transition globally" can be questioned. Besides, this approach is largely framed within the context of the EU's security and geopolitical interests, as indicated by the intention to integrate climate policy implications into external policy considerations, including the Common Security and Defence Policy. The patronizing attitude and limited engagement with the non-EU world underscores a focus on maintaining the EU's stability and defense (read militarization) and securing its financial interests rather than addressing the broader historical and systemic inequities that contribute to global environmental and social challenges (Peteinaki, 2024). What justice, indeed, is being pursued and served?

In *Dismantling Green Colonialism: Energy and Climate Justice in the Arab Region*, Hamza Hamouchene and Katie Sandwell (2023) offer a different characterization of environmental justice (see also Riofrancos, 2025). They argue that genuine environmental justice requires a fundamental transformation of the global economic system; an end to colonial relations that continue to enslave and dispossess people; a break from existing militarist, colonial, and neocolonial projects; consideration of differentiated responsibilities and vulnerabilities between the Global North and South; recognition and repayment of ecological and climate debt to countries in the Global South; attention to geographic contingencies;

technology transfer to empower local communities; promotion of energy democracy and energy sovereignty; and focus on agriculture and food security.

For them, a just transition involves shifting from an economic system centered on excessive resource extraction and human exploitation to one focused on restoring and regenerating territories, as well as upholding people's rights and dignity. A comprehensive and radical vision of a just transition acknowledges that environmental destruction, capitalist extraction, imperialist violence, inequality, exploitation, expropriation, and marginalization based on race, class, and gender are interconnected effects of a single global system. 'Solutions' that attempt to tackle only one aspect, such as environmental crises, without considering the social, cultural, and economic structures that contribute to them, will inevitably fail (Hamouchene, 2023, p. 44).

Too often, though, green transition and discussions on sustainability serve as a superficial cover for neo-colonial practices of exploitation and domination in a world characterized by enforced liberalization, unfair trade agreements, and an imperial quest for influence and energy resources (Hamouchene, 2023, p. 44). Boukhatem (2023, p. 270) notices that EU-centric policies remain to be neocolonial in that they concentrate on facilitating a green transition in Europe "at the expense of cheap resources and undervalued labour in the global South." Similarly, El Nour speaks out against the costs of a just European transition for the global majority:

> In North Africa, historically unequal ecological exchange is intertwined with relationships of exchange with European countries. Here, unequal exchange affects the allocation of water, land, climatic resources, energy and labour power, all of which are geared towards food production for European markets. North African countries bear the environmental costs, as their local ecosystems are destroyed and their natural resources depleted. They also bear the economic costs by generating surplus value through international trade with European countries. This, in turn, has far-reaching consequences for the sustainability of resources, energy and land in North Africa, as well as for the ability to develop frameworks for food sovereignty and to achieve a just transition locally. Unequal environmental exchange perpetuates an imperialist way of life in the capitalist core countries, while severely restricting the chances of a just transition in the South. What is presented as an environmentally and socially just transition for Europe is not necessarily the case for the peripheries attached to the continent in the southern Mediterranean and West Africa. (El Nour, 2023, p. 120)

EU policies may claim to be just for their citizens, but they fall short of being emancipatory for the rest of the world. Instead, they often perpetuate Europe's position as the center and metropole by subordinating and exploiting both non-EU human beings and other material actants.

An essential step towards emancipatory worlding, thus, involves reconceptualizing the relationship between humans and the material and natural worlds. Barca (2024) argues that labor is a gendered and racialized interface between society and nature. She contends that,

> Making space for labor in the politics of socio-ecological revolution … means aiming for a truly democratic, workers-controlled production system, where alienation is actively countered by a collective reappropriation of the products of labor and by a truly democratic decision-making process over the use of the surplus. Such strategy must be based on an extended concept of class relations that goes beyond the wage labour relation, and towards a broader conception of work as a (gendered and racialised) mediator of social metabolism. I conclude that ecosocialist degrowth should take the form of a struggle for dealienating both industrial and meta-industrial labor. (Barca, 2024, p. 17)

Barca highlights the injustice of shifting the costs of environmental cleanups onto labor and emphasizes the rights to compensation and self-determination for working-class communities in a transition to an environmentally sustainable economy. This reconceptualization of the relationship between humans and the more-than-human world inevitably necessitates a rethinking of human-to-human socio-economic relationships. By addressing how labor serves as a critical interface between society and nature, emancipatory worlding also requires re-creating the socio-economic structures that govern human interactions.

Another requirement of emancipatory worlding is the undermining of species supremacy understood as "the ideology that legitimates violence against non-human life as instrumental to human progress" (Barca, 2024, p. 86). Species supremacy is, in fact, a fundamental element of capitalist and industrial modernity, and its dismantling is essential, as the current system could potentially collapse without this ideology.

Emancipatory worlding further requires the rejection of simplified views of human identity as self-contained and human labor as inherently opposed to non-human nature. This involves resisting species supremacy and the resulting degradation of Earth's systems. Opposing species supremacy, Barca (2024) proposes the concept of interspecies commoning. This idea advocates for anti-mastery relations both among humans and between humans and more-than-humans. Interspecies commoning is a non-individualized system of land access where human and more-than-human nature are equally integral to social identity. Barca views interspecies commoning as an onto-epistemology, a mode of reproduction, and a political principle. It is developed through historical agency and connected to the principle of florestania, or "forestzenship," which recognizes the forest and its inhabitants as a polis—a more-than-human community with its own political subjectivity and equal rights.

Barca contends that human emancipation is inseparable from the emancipation of more-than-humans. She envisions the possibility of a world beyond mastery, where people achieve freedom from exploitation and alienation in conjunction with, rather than separate from, more-than-human. "Liberation here assumes an interspecies meaning: there can be no true emancipation in a degraded and threatened environment where earth others are sacrificed" (Barca, 2024, p. 86). Barca sees this alternative mode of humanity as offering the greatest hope for addressing and potentially resolving the global planetary crisis of our time.

As an emancipatory mode of worlding, interspecies commoning aligns with anti-master thinking and activism. It "speaks to today's labor environmentalism in important ways, pointing towards the possibility of thinking of non-human nature as co-constitutive of the labor subject (rather than its 'other'), and organizing workers beyond the wage relation, struggling for radically anti-master relations with the earth" (Barca, 2024, p. 89). This approach challenges traditional labor dynamics and promotes a vision of labor and environmentalism that integrates the more-than-human as a fundamental component of social and ecological identity.

And what drives emancipatory worlding? In the introduction, I discussed the limitations of fear as a motivator. Many people, perhaps under the influence of positive psychology, regard optimism as an effective motivator. However, optimism often functions as a narrative that maintains the status quo, a-politicizing the discourse. As Chaudhary (2024) notes, terms like resisting, absorbing, accommodating, and recovering are socially passive and politically inert, relying on cultivating a belief system that provides a cohesive life narrative. They foster complacency: optimism in a world that is failing, intelligence that views this as the best possible reality, creativity focused on adaptation rather than radical transformation, and a belief system that affirms the current world while preparing individuals for widespread suffering. Emancipatory worlding requires a shift from these passive responses to active, daring, and critical engagement that ruptures, dismantles, and transforms existing structures.

In *Hope Without Optimism*, Terry Eagleton (2015) examines the concept of hope, setting it apart from optimism. He contends that optimism is frequently naive, resting on the assumption that things will naturally improve. By contrast, hope is a more nuanced stance that recognizes challenges and yet continues to endure. Eagleton critiques modern culture's inclination to equate hope with wishful thinking, proposing instead a more grounded and potentially radical perspective on hope—one that acknowledges suffering and struggle as integral to the human experience (see also Gana, 2023; Gros, 2025; Norman & Welchman, 2025; Solnit, 2016; van der Lugt, 2025; Wilderson, 2020).

I argued that even hope can serve as a barrier to meaningful and timely action; therefore, I call for dis-hoping, radical pessimism, undefeated despair understood as "despair without fear, without resignation, without a sense of defeat" (Berger, 2006, p. 604). Rather than fearing an impending apocalypse, we must learn to constantly live in the time of the end. Instead of hoping to preserve the status quo while avoiding further ecological harm, we must endorse the collapse of the current system, while our undefeated despair compels us to take desperate measures. Rather than feasting on and perpetuating relations of mastery including species supremacy, we must embrace and cherish our material entanglements. The task is for the recalcitrant Exhausted of the Earth in all its heterogeneity and contingencies to struggle for nothing short of emancipatory worlding.

## References


Abbasi, K. (2025). The law may be an ass, but is it our last hope to protect health? *BMJ, 389*, r1324. https://doi.org/10.1136/bmj.r1324

Barca, S. (2024). *Workers of the earth: Labour, ecology and reproduction in the age of climate change*. Pluto Press.

Bennett, J. (2010). *Vibrant matter: A political ecology of things.* Duke University Press.

Berger, J. (2006). Undefeated despair. *Critical Inquiry, 32*(4), 602–609. https://doi.org/10.1086/508082

Bosch, T. (2025). Decolonisation is not a vibe: On anti-capitalist praxis, citation politics and epistemic refusal. *Media, Culture and Society, 47*(8), 1714–1724. https://doi.org/10.1177/01634437251360382

Boukhatem, I. (2023). The challenges of the energy transition in fossil fuel exporting countries: The case of Algeria. In H. Hamouchene & K. Sandwell (Eds.), *Dismantling green colonialism: Energy and climate justice in the Arab Region* (1st ed., pp. 252–274). Pluto Press. https://doi.org/10.2307/jj.7583922.18

Burawoy, M. (2025). Why and how should sociologists speak out on Palestine? *The Sociological Review, 73*(2), 249–260. https://doi.org/10.1177/00380261251321676

Chaudhary, A. S. (2024). *The exhausted of the Earth: Politics in a burning world*. Penguin.

Chiew, S., Mayes, E., Maiava, N., Villafaña, D., & Abhayawickrama, N. (2024). Funny climate activism? A collaborative storied analysis of young climate advocates' digital activisms. *Global Studies of Childhood, 14*(4), 390–411. https://doi.org/10.1177/20436106241241338

Dai, M. (2024, December 12). Daily cartoon: Thursday, December 12. *The New Yorker*. https://www.newyorker.com/cartoons/daily-cartoon/thursday-december-12th-which-destinations

Eagleton, T. (2015). *Hope without optimism*. Yale University Press.

Ferdinand, M. (2022). *Decolonial ecology: Thinking from the Caribbean world*. Polity.

Frank, D. J., Hironaka, A., & Schofer, E. (2000). The nation-state and the natural environment over the twentieth century. *American Sociological Review, 65*(1), 96–116. https://doi.org/10.2307/2657291

Gana, N. (2023). *Melancholy acts: Defeat and cultural critique in the Arab world*. Fordham University Press.

Ghosh, A. (2024). *Smoke and ashes: Opium's hidden histories*. John Murray.

Gros, F. (2025). *A philosophy of shame: A revolutionary emotion*. Verso Books.

Hamouchene, H. (2023). The energy transition in North Africa: Neocolonialism again! In H. Hamouchene & K. Sandwell (Eds.), *Dismantling green colonialism: Energy and climate justice in the Arab Region* (1st ed., pp. 29–48). Pluto Press. https://doi.org/10.2307/jj.7583922.7

Heise, U. K. (2016). *Imagining extinction: The cultural meanings of endangered species*. The University of Chicago Press.

Horkheimer, M., & Adorno, T. W. (2016). *Dialectic of enlightenment*. Verso.

Hutcheon, L. (1988). *A poetics of postmodernism: History, theory, fiction*. Routledge.

Lordon, F. (2014). *Willing slaves of capital: Spinoza and Marx on desire*. Verso.

Nixon, R. (2011). *Slow violence and the environmentalism of the poor*. Harvard University Press. https://doi.org/10.2307/j.ctt2jbsgw

Nkrumah, K. (1965). *Neo-colonialism: The last stage of imperialism*. Thomas Nelson & Sons, Ltd. https://www.marxists.org/ebooks/nkrumah/nkrumah-neocolonialism.pdf

Norman, J., & Welchman, A. (2025). The politics of pessimism. In C. Marshall & S. Shapshay (Eds.), *The new Cambridge companion to schopenhauer*. Cambridge University Press.

El Nour, S. (2023). Towards a just agricultural transition in North Africa. In H. Hamouchene & K. Sandwell (Eds.), *Dismantling green colonialism: Energy and climate justice in the Arab Region* (1st ed., pp. 109–133). Pluto Press. https://doi.org/10.2307/jj.7583922.11

Pannett, R. (2025, January 22). Colorado zoo sued over plans to bring elephants from Swaziland. *The Washington Post*. https://www.washingtonpost.com/climate-environment/2025/01/22/colorado-zoo-elephant-lawsuit/

Peteinaki, M. (2024). Reevaluating sustainable development goals and European climate policies through the Doughnut Model. [Master's dissertation, Hellenic Open University]. Apothesis Institutional Repository. https://apothesis.eap.gr/archive/item/212712?lang=en

Riofrancos, T. (2025). *Extraction: The frontiers of green capitalism*. W.W. Norton.

Solnit, R. (2016). *Hope in the dark: Untold histories, wild possibilities* (3rd ed.). Haymarket Books.

Sultana, F. (Ed.). (2024). *Confronting climate coloniality: Decolonizing pathways for climate justice* (1st ed.). Routledge. https://doi.org/10.4324/9781003465973

The Late Show with Stephen Colbert. (2025, September 26). Meanwhile… Central Perk NYC | Disney Scuba Robbery | Crackless Tomatoes | Coffee makes you happy [Video]. YouTube. https://www.youtube.com/watch?v=NDkjBGiUZig

Van der Lugt, M. (2025). *Hopeful pessimism*. Princeton University Press.

Wilderson, F. B. (2020). *Afropessimism*. Liveright.

# References


Abbasi, K. (2025). The law may be an ass, but is it our last hope to protect health? *BMJ, 389*, r1324. https://doi.org/10.1136/bmj.r1324

Abrahamse, W. (2019). *Encouraging pro-environmental behavior: What works, what doesn't, and why*. Academic.

Ackerman, A. (2015). A "fluid" truism: *Dr. Strangelove's* confrontation of the malleability of knowledge and truth. *Young Scholars in Writing, 12*, 75–82. https://youngscholarsinwriting.org/index.php/ysiw/article/view/211

Ahmed, S. (2004). *The cultural politics of emotion*. Routledge.

Ajzen, I. (1985). From intentions to actions: A theory of planned behavior. In J. Kuhl & L. Beckmann (Eds.), *Action control: From cognition to behavior* (pp. 11–39). Springer. https://doi.org/10.1007/978-3-642-69746-3_2

Ajzen, I. (1991). The theory of planned behavior. *Organizational Behavior and Human Decision Processes, 50*(2), 179–211. https://doi.org/10.1016/0749-5978(91)90020-T

Ajzen, I., & Kruglanski, A. W. (2019). Reasoned action in the service of goal pursuit. *Psychological Review, 126*(5), 774–786. https://doi.org/10.1037/rev0000155

Althusser, L. (1970). Ideology and ideological state apparatuses. In *Lenin and philosophy and other essays* (pp. 121–176). Monthly Review Press.

American Psychological Association. (2009). *Psychology and global climate change: Addressing a multifaceted phenomenon and set of challenges*. Retrieved from https://www.apa.org/science/about/publications/climate-change

An Ecomodernist Manifesto. (2015). *The breakthrough institute*. https://thebreakthrough.org/manifesto/manifesto-english





M. Zekavat, *Leveraging Satire for Environmental Advocacy*, https://doi.org/10.1007/978-3-032-04838-7

APA. (2009). *Psychology & global climate change: Addressing a multifaceted phenomenon and set of challenges.* Retrieved April 1, 2022, from, https://www.apa.org/science/about/publications/climate-change

Appadurai, A. (1986). Introduction: Commodities and the politics of value. In A. Appadurai (Ed.), *The social life of things: Commodities in cultural perspective* (pp. 3–63). Cambridge University Press. https://doi.org/10.1017/CBO9780511819582.003

Apter, M. J. (2005). *Personality dynamics: Key concepts in reversal theory.* Apter International.

Apter, M. J. (2013). Developing reversal theory: Some suggestions for future research. *Journal of Motivation, Emotion, and Personality, 1*(1), 1–8. https://doi.org/10.12689/jmep.2013.101

Arendt, H. (1998). *The human condition* (2nd ed.). University of Chicago Press.

Bakhtin, M. (1984). *Rabelais and his world.* MIT Press.

Barca, S. (2020). *Forces of reproduction: Notes for a counter-hegemonic Anthropocene.* Cambridge University Press. https://doi.org/10.1017/9781108768380

Barca, S. (2024). *Workers of the earth: Labour, ecology and reproduction in the age of climate change.* Pluto Press.

Battistoni, A. (2025). *Free gifts: Capitalism and the politics of nature.* Princeton University Press.

Baudrillard, J. (1994). *Simulacra and simulation.* University of Michigan Press.

Baym, G., & Shah, C. (2011). Circulating struggle: The online flow of environmental advocacy clips from the daily show and the Colbert report. *Information, Communication & Society, 14*(7), 1017–1038. https://doi.org/10.1080/1369118X.2011.554573

Beauman, N. (2022). *Venomous lumpsucker.* Soho Press.

Beauman, N. (n.d.). *Ned Beauman.* Retrieved March 25, 2025, from https://www.nedbeauman.co.uk/

Bennett, J. (2010). *Vibrant matter: A political ecology of things.* Duke University Press.

Berger, J. (2006). Undefeated despair. *Critical Inquiry, 32*(4), 602–609. https://doi.org/10.1086/508082

Bergson, H. (1911). *Laughter: An essay on the meaning of the comic.* Macmillan.

Besussi, A. (Ed.). (2017). *A companion to political philosophy: Methods, tools, topics.* Routledge.

Bevis, M. (2013). *Comedy: A very short introduction.* Oxford University Press.

Billig, M. (2018). Positive psychology: Humour and the virtues of negative thinking. In F. Maon, A. Lindgreen, J. Vanhamme, R. J. Angell, & J. Memery (Eds.), *Not all claps and cheers: Humour in business and society relationships* (pp. 3–13). Routledge.

Bloom, E. A., & Bloom, L. D. (1979). *Satire's persuasive voice.* Cornell University Press.

Bogad, L. M. (2016). *Tactical performance: The theory and practice of serious play.* Routledge.

Böhmermann, J. (2021). Jetzt schon Wahlsieger: Das Auto! | ZDF Magazin Royale. YouTube. https://youtu.be/ybTHWzmlw70

Bok, H. (2018). Baron de Montesquieu, Charles-Louis de Secondat. In E. N. Zalta (Ed.), *The Stanford encyclopedia of philosophy.* Plato. https://plato.stanford.edu/archives/win2018/entries/montesquieu/

Bonneuil, C., & Fressoz, J.-B. (2016). *The shock of the Anthropocene: The earth, history and us.* Verso Books.

Bore, I.-L., & Reid, G. (2014). Laughing in the face of climate change? Satire as a device for engaging audiences in public debate. *Science Communication, 36*(4), 454–478. https://doi.org/10.1177/1075547014534076

Borum, V. (2023). *The revolution will be hilarious: Comedy for social change and civic power.* New York University Press.

Bosch, T. (2025). Decolonisation is not a vibe: On anti-capitalist praxis, citation politics and epistemic refusal. *Media, Culture and Society, 47*(8), 1714–1724. https://doi.org/10.1177/01634437251360382

Boukes, M. (2019). Infotainment. In T. P. Vos, F. Hanusch, D. Dimitrakopoulou, M. Geertsema-Sligh, & A. Sehl (Eds.), *The international encyclopedia of journalism studies* (pp. 1–9). John Wiley & Sons, Inc.. https://doi.org/10.1002/9781118841570.iejs0132

Boukhatem, I. (2023). The challenges of the energy transition in fossil fuel exporting countries: The case of Algeria. In H. Hamouchene & K. Sandwell (Eds.), *Dismantling green colonialism: Energy and climate justice in the Arab Region* (1st ed., pp. 252–274). Pluto Press. https://doi.org/10.2307/jj.7583922.18

Boykoff, M. (2019). *Creative (climate) communications: Productive pathways for science, policy, and society.* Cambridge University Press. https://doi.org/10.1017/9781108164047

Boykoff, M., & Osnes, B. (2019). A laughing matter? Confronting climate change with humor. *Political Geography, 68*, 154–163. https://doi.org/10.1016/j.polgeo.2018.09.006

Branch, M. P. (2014). Are you serious? A modest proposal for environmental humor. In G. Garrard (Ed.), *The Oxford handbook of ecocriticism* (pp. 378–391). Oxford University Press. https://doi.org/10.1093/oxfordhb/9780199742929.013.021

Brenner, J., & Fraser, N. (2017). What is progressive neoliberalism?: A debate. *Dissent, 64*(2), 130. https://doi.org/10.1353/dss.2017.0051

Bresnihan, P., & Millner, N. (2023). *All we want is the Earth: Land, labour and movements beyond environmentalism.* Bristol University Press.

Brick, C., & Lewis, G. L. (2016). Unearthing the "green" personality: Core traits predict environmentally friendly behavior. *Environment and Behavior, 48*(5), 635–658. https://doi.org/10.1177/0013916514554695

Broderick, M. (1993). Surviving armageddon: Beyond the imagination of disaster. *Science Fiction Studies, 20*(3), 362–382. http://www.jstor.org/stable/4240277.

Broderick, M. (2017). *Reconstructing strangelove: Inside Stanley Kubrick's nightmare comedy*. Columbia University Press. https://doi.org/10.7312/brod17708

Brody, R. (2022, January 6). The crude demagogy of don't look up: Adam McKay's star-studded allegory of climate change has a cynically apolitical view of politics. *The New Yorker*. https://www.newyorker.com/culture/the-front-row/the-crude-demagogy-of-dont-look-up

Buck, H. J. (2019). *After geoengineering: Climate tragedy, repair, and restoration*. Verso Books.

Buell, F. (2003). *From apocalypse to way of life: Environmental crisis in the American century*. Routledge.

Buell, L. (1995). *The environmental imagination: Thoreau, nature writing, and the formation of American culture*. Harvard University Press.

Burawoy, M. (2025a). Palestine through a South African lens. *New Left Review, 153*, 133–145. https://newleftreview.org/issues/ii153/articles/michael-burawoy-palestine-through-a-south-african-lens

Burawoy, M. (2025b). Why and how should sociologists speak out on Palestine? *The Sociological Review, 73*(2), 249–260. https://doi.org/10.1177/00380261251321676

Butler, J. (2004). *Precarious life: The powers of mourning and violence*. Verso.

Butler, J. (2009). *Frames of war: When is life grievable?* Verso.

Butler, J. (2024). *Who's afraid of gender?* Farrar, Straus and Giroux.

CAConrad. (2015). Slaves of hope live only for tomorrow. *Poetry Foundation*. https://www.poetryfoundation.org/poems/58067/slaves-of-hope-live-only-for-tomorrow

Campbell, C., & Abdullah, T. (2023). *Race, representation, and satire*. Lexington Books.

Caracciolo, M. (2022). *Contemporary fiction and climate uncertainty: Narrating unstable futures*. Bloomsbury Academic.

Caracciolo, M., Crockford, S., Lambert, S., & Ulstein, G. (2021). Phenomenology of everyday climate: An ethnographic approach to metaphor, affect, and the nonhuman. *ISLE Interdisciplinary Studies in Literature and Environment, 30*(2), 353–376.

Caradonna, J. L., et al. (2015, May 6). *A degrowth response to An ecomodernist manifesto*. Resilience.org. https://www.resilience.org/stories/2015-05-06/a-degrowth-response-to-an-ecomodernist-manifesto/

Carpenter, C. J. (2020). Elaboration likelihood model. In *The international encyclopedia of media psychology* (pp. 1–14). Wiley.

Carton, W., & Malm, A. (2024). *Overshoot: How the world surrendered to climate breakdown*. Verso Books.

Carton, W., & Malm, A. (2025). *The long heat: Climate politics when it's too late*. Verso.

Chandler, K. R. (2014). Poisonwood persuasion: Rhetorical roles of humor in environmental literature. *Interdisciplinary Literary Studies, 16*(2), 326–347. https://muse.jhu.edu/article/554074

Chattoo, C. B., & Feldman, L. (2020). *A comedian and an activist walk into a bar: The serious role of comedy in social justice*. University of California Press.

Chaudhary, A. S. (2024). *The exhausted of the Earth: Politics in a burning world*. Penguin.

Cheng, P., Wang, W., & Yang, S. (2024). Doing the right thing: How to persuade travelers to adopt pro-environmental behaviors? An elaboration likelihood model perspective. *Journal of Hospitality and Tourism Management, 59*, 191–209.

Chiew, S., Mayes, E., Maiava, N., Villafaña, D., & Abhayawickrama, N. (2024). Funny climate activism? A collaborative storied analysis of young climate advocates' digital activisms. *Global Studies of Childhood, 14*(4), 390–411. https://doi.org/10.1177/20436106241241338

Clan, W.-T. (1993). *Wu-Tang Clan ain't nuthing ta f'wit [Song]. On Enter the Wu-Tang (36 Chambers) [Album]*. Loud Records.

Clean Energy Regulator. (n.d.). *The safeguard mechanism*. Retrieved from https://www.cleanenergyregulator.gov.au/NGER/The-Safeguard-Mechanism

Climate Change News. (2022, September 8). *Australia passes climate law targeting 43% emissions cuts by 2030*. Retrieved from https://www.climatechangenews.com/2022/09/08/australia-passes-climate-law-targeting-43-emissions-cuts-by-2030/

Colbert, S. (2021a). *Aging politicians and meat eaters stand in the way of America's climate progress*. YouTube. https://youtu.be/YcFlDnxiBiM

Colbert, S. (2021b). *Americans having less sex, avoiding movie theaters as U.S. braces for possible fifth wave*. YouTube. https://youtu.be/moJJuJd5S4Q

Colbert, S. (2021c). *British royals can't stop talking about Biden's alleged gas leak | Pfizer may have a Covid cure*. YouTube. https://youtu.be/MkopjQw0VBs

Colbert, S. (2021d). *Climate: Changed - Is the sun god angry at us?* YouTube. https://youtu.be/2I_K63hOgzk

Colbert, S. (2021e). *Food and gas prices spike, and Rep. Mo Brooks wore body armor at 'Stop the Steal' rally*. YouTube. https://youtu.be/RPrTiLtqXVM

Colbert, S. (2021f). *Is systemic climate action possible? Stephen seeks guidance from Mama Earth*. YouTube. https://youtu.be/rbzMvJf1Vxg

Colbert, S. (2021g). *Queen urges leaders to rise above climate politics | Preparing kids for the vaccine*. YouTube. https://youtu.be/meUe2Vg4pOo

Colbert, S. (2021h). *Rainfall swamps the west, and GOP members conspired with Jan. 6th riot organizers*. YouTube. https://youtu.be/TiBtWFOLLjw

Colbert, S. (2021i). *Stephen Colbert's Dil-Do's & Dil-Don'ts*. YouTube. https://youtu.be/Pzgtt_QkIcE

Colbert, S. (2021j). *U.S. warns Russia not to take Ukraine | Earth's black box will record humanity's demise*. YouTube. https://youtu.be/QjEEiPGs9YE

Colbert, S. (2021k). *Uh-oh, water is now a commodity - Stephen Colbert's most unfortunate segment*. YouTube. https://youtu.be/gnEDF0tTReI

Colbert, S. (2021l). *World leaders try to wish away climate change, Bezos skips Halloween to party with Bill Gates*. YouTube. https://youtu.be/-b0G_CCaTI4

Colbert, S. (2022). *Meanwhile... Elon Musk and Grimes welcome another child | Metallica's Lars Ulrich is now a toilet*. YouTube. https://youtu.be/q8zlh8XKfLc

Condren, C., Milner Davis, J., McCausland, S., & Phiddian, R. (2008). Defining parody and satire: Australian copyright law and its new exception: Part 2—Advancing ordinary definitions. *Media and Arts Law Review, 13*(4), 401–421.

Cooke, E. F. (2007). Understanding the enemy: The dialogue of fear in *Fear and Desire* and *Dr. Strangelove*. In J. J. Abrams (Ed.), *The philosophy of Stanley Kubrick* (pp. 9–31). The University Press of Kentucky.

Corral-Verdugo, V., Frías-Armenta, M., Tapia-Fonllem, C. O., & Fraijo-Sing, B. S. (2012). Protecting natural resources: Psychological and contextual determinants of freshwater conservation. In S. D. Clayton (Ed.), *The Oxford handbook of en-vironmental and conservation psychology* (pp. 581–597). Oxford University Press. https://doi.org/10.1093/oxfordhb/9780199733026.013.0030

Coulson, S. (2007). Funnier than unhappiness: Adorno and the art of laughter. *New German Critique, 100*, 141–163. http://www.jstor.org/stable/27669190

Crist, E. (2016). The reaches of freedom: A response to an ecomodernist manifesto. *Environmental Humanities, 7*(1), 245–254. https://doi.org/10.1215/22011919-3616452

Critchley, S. (2002). *On humour*. Routledge.

Cummins, A. (2022, July 17). Ned beauman's *Venomous Lumpsucker* mischievous review. *The Guardian*. https://www.theguardian.com/books/2022/jul/17/venomous-lumpsucker-by-ned-beauman-review-mischievous-meaning-of-life-satire

Dai, M. (2024, December 12). Daily cartoon: Thursday, December 12. *The New Yorker*. https://www.newyorker.com/cartoons/daily-cartoon/thursday-december-12th-which-destinations

Dalgleish, T. (2004). The emotional brain. *Nature Reviews. Neuroscience, 5*(7), 583–589. https://doi.org/10.1038/nrn1432

Davidson, H. (2017, August 2). Emissions from NT's McArthur basin would dwarf those from Adani coalmine. *The Guardian*. https://www.theguardian.

com/environment/2017/aug/02/emissions-from-nts-mcarthur-basin-would-dwarf-those-from-adanicoalmine

De Groot, J. I. M., & Thøgersen, J. (2013). Values and pro-environmental behavior. In L. Steg, A. E. van den Berg, & J. I. de Groot (Eds.), *Environmental psychology: An introduction* (pp. 141–152). BPS Blackwell.

De Groot, J. I. M., & Thøgersen, J. (2018). Values and pro-environmental behaviour. In L. Steg & J. de Groot (Eds.), *Environmental psychology* (pp. 167–178). Blackwell. https://doi.org/10.1002/9781119241072.ch17

De Leeuw, A., Valois, P., Ajzen, I., & Schmidt, P. (2015). Using the theory of planned behavior to identify key beliefs underlying pro-environmental behavior in high- school students: Implications for educational interventions. *Journal of Environ-mental Psychology, 42*, 128–138. https://doi.org/10.1016/j.jenvp.2015.03.005

de Waal, F. B. M. (1997, July). Are we in anthropodenial? In *Discover magazine* (pp. 50–53).

Derrida, J. (1992). Before the law. In D. Attridge (Ed.), *Acts of literature* (pp. 181–220). Routledge.

Deutsch, J. I. (2016). 'No communist could dream of a more effective anti-American film': Dr. Strangelove and its red reviewers. *Historical Journal of Film, Radio and Television, 36*(1), 40–52. https://doi.org/10.1080/01439685.2015.1134108

DiCaglio, J. (2015). Ironic ecology. *Interdisciplinary Studies in Literature and Environment, 22*(3), 447–465. https://doi.org/10.1093/isle/isu106

Douglas, R. (2025). *Introduction: Living and wasting time in the world-system. Women in world-literature*. Liverpool University Press.

Doyle, J. (2022). Communicating climate change in 'Don't Look Up'. *Journal of Science Communication, 21*(05), C02. https://doi.org/10.22323/2.21050302

*Dr. Strangelove or: How I Learned to Stop Worrying and Love the Bomb*. (1964). [Film]. Directed by S. Kubrick. Columbia Pictures. https://archive.org/details/dr.-strangelove-1964

Dr. Strangelove: A Continuity Transcript. (n.d.). University of Notre Dame. https://www3.nd.edu/~dlindley/handouts/DrS.htm

Eagleton, T. (2015). *Hope without optimism*. Yale University Press.

Eaton, M. (2012). Dark comedy from *Dr. Strangelove* to the dude. In A. Horton & J. E. Rapf (Eds.), *A companion to film comedy* (pp. 315–339). Wiley. https://doi.org/10.1002/9781118327821.ch15

Ehrenreich, B. (2009). *Smile or die: How positive thinking fooled America and the world*. Granta Books.

El Nour, S. (2023). Towards a just agricultural transition in North Africa. In H. Hamouchene & K. Sandwell (Eds.), *Dismantling green colonialism: Energy and climate justice in the Arab Region* (1st ed., pp. 109–133). Pluto Press. https://doi.org/10.2307/jj.7583922.11

Elliott, R. C. (1960). *The power of satire: Magic, ritual, art.* Princeton University Press.

Estlund, D. (Ed.). (2012). *The Oxford handbook of political philosophy.* Oxford University Press.

European Commission. (2019). *The European green deal (COM/2019/640 final).*https://eur-lex.europa.eu/legal-content/EN/TXT/?uri=CELEX%3A52019DC0640

Evans, D. (1996). *An introductory dictionary of Lacanian psychoanalysis.* Routledge.

Feldman, L. (2017). Assumptions about science in satirical news and late-night comedy. In K. H. Jamieson, D. Kahan, & D. A. Scheufele (Eds.), *The Oxford handbook of the science of science communication* (pp. 321–332). Oxford University Press. https://doi.org/10.1093/oxfordhb/9780190497620.013.35

Feldman, L., Leiserowitz, A., & Maibach, E. (2011). The science of satire: The daily show and the colbert report as sources of public attention to science and the environment. In A. Amarasingam (Ed.), *The Stewart/Colbert effect: Essays on the real impacts of fake news* (pp. 25–46). McFarland.

Feng, G. C., Luo, Y., Yu, Z., & Wen, J. (2023). Effects of rhetorical devices on audience responses with online videos: An augmented elaboration likelihood model. *PLoS One, 18*(3), e0282663.

Ferdinand, M. (2022). *Decolonial ecology: Thinking from the Caribbean world.* Polity.

Fiennes, S. (2006). *The pervert's guide to cinema: 1, 2, 3* [Film]. Lone Star, Mischief Films, Amoeba Film.

Foster, J. B. (1999). Marx's theory of metabolic rift: Classical foundations for environmental sociology. *American Journal of Sociology, 105*(2), 366–405. https://doi.org/10.1086/210315

Foster, J. B. (2000). *Marx's ecology: Materialism and nature.* Monthly Review Press.

Foster, J. B. (2013). Marx and the rift in the universal metabolism of nature. *Monthly Review, 65*(7), 12. https://monthlyreview.org/2013/12/01/marx-rift-universal-metabolism-nature/

Foucault, M. (1975). *Discipline and punish: The birth of the prison.* [English edition published in 1977]. Pantheon Books.

Foucault, M. (1977). *Discipline and punish: The birth of the prison.* Pantheon Books.

Foucault, M. (1991). *Governmentality.* In G. Burchell, C. Gordon, & P. Miller (Eds.), *The Foucault effect: Studies in governmentality* (pp. 87–104). University of Chicago Press.

Frank, D. J., Hironaka, A., & Schofer, E. (2000). The Nation-State and the Natural Environment over the Twentieth Century. *American Sociological Review, 65*(1), 96–116. https://doi.org/10.2307/2657291

Fraser, N. (2022). *Cannibal capitalism: How our system is devouring democracy, care, and the planet—and what we can do about it*. Verso.

Fraser, N. (2016). Expropriation and exploitation in racialized capitalism: A reply to Michael Dawson. *Critical Historical Studies, 3*(1), 163–178. https://doi.org/10.1086/685814

Fressoz, J.-B. (2024). *More, more, more: A history of the world's excesses*. Penguin Books.

Friedman, H. H., & Friedman, L. W. (2020). The pen is mightier than the sword: Humor as a social justice tool. *Review of Contemporary Philosophy, 19*, 26–42. https://doi.org/10.22381/RCP1920202

Gana, N. (2023). *Melancholy acts: Defeat and cultural critique in the Arab world*. Fordham University Press.

García Mainar, L. M. (1994). Kubrick's *Dr. Strangelove*: Desire and the fragmenting of character. *Revista Alicantina de Estudios Ingleses, 7*, 67. https://doi.org/10.14198/raei.1994.7.06

Garrard, G. (2023). *Ecocriticism* (3rd ed.). Routledge.

Gatersleben, B., & Steg, L. (2013). Affective and symbolic aspects of environmental behavior. In L. Steg, A. E. van den Berg, & J. I. de Groot (Eds.), *Environmental psychology: An introduction* (pp. 165–174). BPS Blackwell.

Ghosh, A. (2016). *The great derangement: Climate change and the unthinkable*. University of Chicago Press.

Ghosh, A. (2021). *The nutmeg's curse: Parables for a planet in crisis*. University of Chicago Press.

Ghosh, A. (2024). *Smoke and ashes: Opium's hidden histories*. John Murray.

Giamario, P. (2023). *Laughter as Politics: Critical Theory in an Age of Hilarity*. Edinburgh University Press.

Giladi, T. (2024). The emergence of Marx's concept of subsumption. *Archiv für Geschichte der Philosophie, 106*(3), 611–631. https://doi.org/10.1515/agph-2022-0074

Griffin, D. (1994). *Satire: A critical reintroduction*. University Press of Kentucky.

Grimbeek, M. (2017). *Margaret Atwood's environmentalism: Apocalypse and sat-ire in the MaddAddam trilogy*. Karlstad University Studies.

Gros, F. (2025). *A philosophy of shame: A revolutionary emotion*. Verso Books.

Guo, X., Xu, B., Yu, H., Burnett, W. C., Li, S., Lian, E., Zhu, Z., Zhao, S., Chen, G., Duan, X., Dimova, N., Wang, Y., Zhuang, G., & Yu, Z. (2022). Exploration of deep ocean ferromanganese nodule fields using radon as a tracer. *Geophysical Research Letters, 49*(22), 100726. https://doi.org/10.1029/2022gl100726

Haider, L. J., & Rieser, J. P. S. (2025). From reflexive co-production to diffractive co-becoming: Insights from new materialism for sustainability sciences. *Humanities and Social Sciences Communications, 12*(1), 787. https://doi.org/10.1057/s41599-025-05101-6

Hamilton, C. (2010). *Requiem for a species: Why we resist the truth about climate change*. Earthscan.

Hamouchene, H. (2023). The energy transition in North Africa: Neocolonialism again! In H. Hamouchene & K. Sandwell (Eds.), *Dismantling green colonialism: Energy and climate justice in the Arab Region* (1st ed., pp. 29–48). Pluto Press. https://doi.org/10.2307/jj.7583922.7

Hamouchene, H., & Sandwell, K. (Eds.). (2023). *Dismantling green colonialism: Energy and climate justice in the Arab region* (1st ed.). Pluto Press. https://doi.org/10.2307/jj.7583922

Haraway, D. (2015). Anthropocene, capitalocene, plantationocene, chthulucene: Making kin. *Environmental Humanities, 6*(1), 159–165. https://doi.org/10.1215/22011919-3615934

Haraway, D. (2016). *Staying with the trouble: Making kin in the Chthulucene*. Duke University Press.

Haraway, D. J. (1985). A cyborg manifesto: Science, technology, and socialist-feminism in the late twentieth century. *Socialist Review, 80*, 65–108.

Hardin, G. (1968). The tragedy of the commons. *Science, 162*(3859), 1243–1248. https://doi.org/10.1126/science.162.3859.1243

Hardt, M., & Negri, A. (2000). *Empire*. Harvard University Press.

Hardt, M., & Negri, A. (2004). *Multitude: War and democracy in the age of Empire*. Penguin Press.

Hardt, M., & Negri, A. (2009). *Commonwealth*. Harvard University Press.

Harvey, D. (2004). "The 'new' imperialism: accumulation by dispossession". *Socialist Register. 40*: 63–87.

Hedhli, K. E., & Zourrig, H. (2023). Dual routes or a one-way to persuasion? The elaboration likelihood model versus the unimodel. *Journal of Marketing Communications, 29*(5), 433–454. https://doi.org/10.1080/13527266.2022.2034033

Heise, U. K. (2016). *Imagining extinction: The cultural meanings of endangered species*. The University of Chicago Press.

Heise, U. K. (2020). The planetary futures of eco-apocalypse. In J. Hay (Ed.), *Apocalypse in American literature and culture* (pp. 268–280). Cambridge University Press. https://doi.org/10.1017/9781108663557.021

Hellmann, O. (2025). Can *Rick and Morty* save the planet? Re-politicizing climate change through humor and animation. *Television and New Media, 26*(3), 336–354. https://doi.org/10.1177/15274764241271200

Hendrix, C. S., Koubi, V., Selby, J., et al. (2023). Climate change and conflict. *Nature Reviews Earth and Environment, 4*, 144–148. https://doi.org/10.1038/s43017-022-00382-w

Heron, K. (2022, September 7). The great unfettering. *New Left Review*. https://newleftreview.org/sidecar/posts/the-great-unfettering

Higgins, S. (2017). Purity of essence in the cold war: *Dr. Strangelove*, paranoia, and bodily boundaries. *Textual Practice, 32*(5), 799–820. https://doi.org/10.1080/0950236x.2016.1277549

Holmes, D., & Hall, S. (2019). *A literature review of best practice communication of climate science and impacts: Guide for policy makers*. Monash Climate Change Communication Research Hub.

*Holy Bible: King James Version*. (1769/2017). Cambridge University Press. https://www.kingjamesbibleonline.org/

Horkheimer, M., & Adorno, T. W. (2002). *Dialectic of enlightenment: Philosophical fragments* (G. Schmid Noerr, Ed.; E. Jephcott, Trans.). Stanford University Press.

Horkheimer, M., & Adorno, T. W. (2016). *Dialectic of enlightenment*. Verso.

Huber. (2022, September 7). Mish-mash ecologism. *New Left Review*. https://newleftreview.org/sidecar/posts/mish-mash-ecologism

Hutcheon, L. (1988). *A poetics of postmodernism: History, theory, fiction*. Routledge.

Independent Australia. (n.d.). *Morrison courts Pentecostal world domination*. https://independentaustralia.net/politics/politics-display/morrison-courts-pentecostal-world-domination,16582

James, E. (2024). Fiction, belief, and climate change: Paratexts, skeptics, and objects of care. In *The Routledge handbook of fiction and belief* (pp. 323–335). Routledge. https://doi.org/10.4324/9781003119456-28

Jennings, B. (2016). Reconceptualizing autonomy: A relational turn in bioethics. *Hastings Center Report, 46*, 11–16. https://doi.org/10.1002/hast.544

Juice Media. (2016a). Honest Government Ad | Adani Coal Mine [Video]. YouTube. https://youtu.be/n_C8S4Bz91M

Juice Media. (2016b). Honest Government Ad | Dakota Access Pipeline [Video]. YouTube. https://youtu.be/a9TR9G5bd7w

Juice Media. (2017a). Honest Government Ad | Adani [Video]. YouTube. https://youtu.be/G1SdCNdPDL8

Juice Media. (2017b). Honest Government Ad | The Great Australian Bight [Video]. YouTube. https://youtu.be/rNUxQ-9AZss?si=oVscXxHNx9vp-fPv

Juice Media. (2018a). Honest Government Ad | Renewable Energy [Video]. YouTube. https://youtu.be/sitPeRlTdNs?si=aW76QdWp28oxDeRV

Juice Media. (2018b). Honest Government Ad | Visit the Northern Territory! [Video]. YouTube. https://youtu.be/cuQZ6rALub8?si=6bJQMYGSA7O-LKVm

Juice Media. (2019a). Honest Government Ad | Climate Emergency & School Strikes [Video]. YouTube. https://youtu.be/OL8a1YEhk_o

Juice Media. (2019b). Honest Government Ad | Preferential Voting [Video]. YouTube. https://youtu.be/bleyX4oMCgM

Juice Media. (2019c). Honest Government Ad | We're Fine [Video]. YouTube. https://youtu.be/cOmdkN6MOwU

Juice Media. (2020a). Honest Government Ad | Coronavirus: Flatten The Curve [Video]. YouTube. https://youtu.be/Hks6Nq7g6P4

Juice Media. (2020b). Honest Government Ad | Gas-led Recovery [Video]. YouTube. https://youtu.be/7CMUtKWubxg

Juice Media. (2020c). Honest Government Ad | Kyoto Carryover Credits [Video]. YouTube. https://youtu.be/92t8np88fEI

Juice Media. (2020d). Honest Government Ad | The Fires [Video]. YouTube. https://youtu.be/6BmbvTvFQ3g?si=23FPAkgLH4dBymVi

Juice Media. (2021a). Honest Government Ad | Carbon Capture & Storage [Video]. YouTube. https://youtu.be/MSZgoFyuHC8

Juice Media. (2021b). Honest Government Ad | COP26 Climate Summit [Video]. YouTube. https://youtu.be/QIyKmqEdgR4

Juice Media. (2021c). Honest Government Ad | Net Zero (feat. Greta Thunberg) [Video]. YouTube. https://youtu.be/1FqXTCvDLeo

Juice Media. (2022a). Honest Government Ad | Carbon Credits & Offsets [Video]. YouTube. https://youtu.be/iCRDseUEEsg

Juice Media. (2022b). Honest Government Ad | GOV.UK [Video]. YouTube. https://youtu.be/qyt3Op2dTc0

Juice Media. (2022c). Honest Government Ad | The Supreme Court (feat. Pixelsmixel) [Video]. YouTube. https://youtu.be/jL5LaLT2BJM

Juice Media. (2022d). Honest Government Ad | Visit Brazil! [Video]. YouTube. https://youtu.be/sIdwf717p0w

Juice Media. (2022e). Honest Government Ad | Visit WA! [Video]. YouTube. https://youtu.be/Key8y1yg2yQ

Juice Media. (2023a). Honest Government Ad | Anti Protest Laws (SA) [Video]. YouTube. https://youtu.be/kecnSHmznic

Juice Media. (2023b). Honest Government Ad | COP31 [Video]. YouTube. https://youtu.be/YBjyxA-wYS4

Juice Media. (2023c). Honest Government Ad | Labor [Video]. YouTube. https://youtu.be/fHB0vDhdM3c

Juice Media. (2023d). Honest Government Ad | the Safeguard Mechanism [Video]. YouTube. https://youtu.be/YrkE_VaMD4k

Juice Media. (2023e). Honest Government Ad | Visit New South Wales! [Video]. YouTube. https://youtu.be/dcqA5Sq8Dqw

Kafka, F. (1995). Before the law. In N. N. Glatzer (Ed.), *Franz Kafka: The complete stories*. Schocken Books.

Kahneman, D. (2011). *Thinking, fast and slow*. Penguin.

Kals, E., & Müller, M. M. (2012). Emotions and environment. In S. D. Clayton (Ed.), *The Oxford handbook of environmental and conservation psychology* (pp. 128–147). Oxford University Press. https://doi.org/10.1093/oxfordhb/9780199733026.013.0008

Kirshner, J. (2001). Subverting the cold war in the 1960s: Dr. Strangelove, the manchurian candidate, and the planet of the apes. *Film & History: An Interdisciplinary Journal of Film and Television Studies, 31*(2), 40–44. https://doi.org/10.1353/flm.2001.a400700

Knowlton, N. (2017). Doom and gloom won't save the world. *Nature, 541*, 271. https://doi.org/10.1038/544271a

Kolker, R. P. (2017). *The extraordinary image: Orson Welles, Alfred Hitchcock, Stanley Kubrick, and the reimagining of cinema*. Rutgers University Press. http://www.jstor.org/stable/j.ctt1ht4v39

Kormos, C., & Gifford, R. (2014). The validity of self-report measures of proenvironmental behaviour: A meta-analytic review. *Journal of Environmental Psychology, 40*, 359–371. https://doi.org/10.1016/j.jenvp.2014.09.003

Kubrick, S. (1964). *Dr. Strangelove or: How I learned to stop worrying and love the bomb*. Columbia Pictures.

Lacan, J. (2006). *Écrits: The first complete edition in English*. W. W. Norton & Company.

Latour, B. (1999). *Pandora's hope: Essays on the reality of science studies*. Harvard University Press.

Latour, B. (2014). Agency at the time of the Anthropocene. *New Literary History, 45*(1), 1–18. https://doi.org/10.1353/nlh.2014.0003

Latour, B. (2015, June). Fifty shades of green: Presentation to the panel on modernism at the Breakthrough Dialog, Sausalito. *Bruno Latour Official Website*.http://www.bruno-latour.fr/sites/default/files/downloads/00-BREAKTHROUGH-06-15_0.pdf

Latour, B. (2017). *Facing Gaia: Eight lectures on the new climatic regime*. Polity.

Latour, B. (2018). *Down to Earth: Politics in the new climatic regime*. Polity.

Lemos, M. (2012). Dr. Jeems Joker or: How I learned to stop worrying and love *Finnegans Wake*. In M. A. Alonso (Ed.), *Weaving new perspectives together: Some reflections on literary studies* (pp. 74–93). Cambridge Scholars Publishing.

Levinas, E. (1969). *Totality and infinity: An essay on exteriority*. Duquesne University Press.

Levinas, E. (1985). *Ethics and infinity: Conversations with Philippe Nemo*. Duquesne University Press.

Levinas, E. (1998). *Otherwise than being, or beyond essence*. Duquesne University Press.

Linden, G. W. (1977). *Dr. Strangelove* and erotic displacement. *Journal of Aesthetic Education, 11*(1), 63–83. https://doi.org/10.2307/3331863

Lindenberg, S. (2018). How cues in the environment affect normative behavior. In L. Steg & J. de Groot (Eds.), *Environmental psychology: An introduction* (2nd ed., pp. 144–153). John Wiley & Sons Ltd.. https://doi.org/10.1002/9781119241072.ch15

Lindenberg, S., & Steg, L. (2007). Normative, gain and hedonic goal frames guiding environmental behavior. *Journal of Social Issues, 63*(1), 117–137. https://doi.org/10.1111/j.1540-4560.2007.00499.x

Lindley, D. (2001). What I learned since I stopped worrying and studied the movie: A teaching guide to Stanley Kubrick's *Dr. Strangelove*. *PS: Political Science and Politics, 34*(3), 663–667. http://www.jstor.org/stable/1353558

Little, H. (2022). The use of satire to communicate science in Don't look up. *Journal of Science Communication, 21*(05), C06. https://doi.org/10.22323/2.21050306

Lordon, F. (2014). *Willing slaves of capital: Spinoza and Marx on desire*. Verso.

Lyytimäki, J. (2021). Ecological crisis as a laughing matter. Uses of humor in environmental communication. In B. Takahashi, J. Metag, J. Thaker, & S. E. Comfort (Eds.), *The handbook of international trends in environmental communication* (pp. 464–478). Routledge. https://doi.org/10.4324/9780367275204-35

Maguire, L. (2018). The scientific world faced with the apocalypse of its own creation: A comparison of *Dr. Strangelove* and *Nine Days of One Year*. *The Journal of Popular Culture, 51*(4), 956–971. https://doi.org/10.1111/jpcu.12704

Malafouris, L. (2013). *How things shape the mind: A theory of material engagement*. MIT Press.

Maland, C. (1979). *Dr. Strangelove* (1964): Nightmare comedy and the ideology of liberal consensus. *American Quarterly, 31*(5), 697–717. https://doi.org/10.2307/2712432

Malecki, W. P., Schneider-Mayerson, M., Petterson, A., Dobrowolska, M., & Thaker, J. (2025). The role of hope and fear in the impact of climate fiction on climate action intentions: Evidence from India and USA. *Poetics, 108*, 101960. https://doi.org/10.1016/j.poetic.2024.101960

Malm, A. (2024). . Verso Books. https://www.versobooks.com/blogs/news/the-destruction-of-palestine-is-the-destruction-of-the-earth

Malm, A. (2025). *The Destruction of Palestine Is the Destruction of the Earth*. United Kingdom: Verso.

Malm, A., & Carton, W. (2024). *Overshoot: How the world surrendered to climate breakdown*. Verso.

Malm, A., & Zetkin Collective. (2021). *White skin, black fuel: On the danger of fossil fascism*. Verso.

Maloney, S. M. (2020). *Deconstructing Dr. Strangelove: The secret history of nuclear war films*. Project MUSE. https://muse.jhu.edu/book/74964

Malthus, T. R. (1798). *An essay on the principle of population: As it affects the future improvement of society, with remarks on the speculations of Mr. Godwin, M. Condorcet, and other writers*. J. Johnson. Retrieved from http://www.esp.org/books/malthus/population/malthus.pdf

Martin, S. (2021, April 26). Scott Morrison tells Christian conference he was called to do God's work as prime minister. *The Guardian*. https://www.theguardian.com/australia-news/2021/apr/26/scott-morrison-tells-christian-conference-he-was-called-to-do-gods-work-as-prime-minister

Martin, R., & Ford, T. (2018). *The psychology of humor: An integrative approach* (2nd ed.). Academic.

Martin, S. (2021, April 26). Scott Morrison tells Christian conference he was called to do God's work as prime minister. *The Guardian*. Retrieved from https://www.theguardian.com/australia-news/2021/apr/26/scott-morrison-tells-christian-conference-he-was-called-to-do-gods-work-as-prime-minister

Marx, K. (1976). *Capital. A critique of political economy*. Penguin.

Marx, N. (2022). Home economics: Sitcom capitalism, conservative comedy, and media conglomeration in post-network television. *Communication, Culture & Critique, 15*(1), 21–35. https://doi.org/10.1093/ccc/tcab065

Mason, M., Fakher Eldin, M., & Dajani, M. (Eds.). (2021). *The untold story of the Golan Heights: Occupation, colonization and Jawlani resistance* (1st ed.). I.B. Tauris.

Massumi, B. (2002). *Parables for the virtual: Movement, affect, sensation*. Duke University Press.

Mayes, E., & Center, E. (2025). Learning with student climate strikers' humour: Towards critical affective climate justice literacies. In C. Russell, P. Chandler, & J. Dillon (Eds.), *Humour and environmental education* (pp. 30–48). Routledge. https://doi.org/10.4324/9781003582120-3

McBrien, J. (2016). Accumulating extinction: Planetary catastrophism in the Necrocene. In J. W. Moore (Ed.), *Anthropocene or capitalocene? Nature, history, and the crisis of capitalism* (pp. 116–137). PM Press.

McDonald, F. (2020). The last laughs of doomsday humor. In *Apocalypse in American literature and culture* (pp. 281–292). Cambridge University Press. https://doi.org/10.1017/9781108663557.022

McDonald, F. (2022). *Posthumorism: The modernist affect of laughter*. Bloomsbury Academic.

McKay, A. (2021). *Don't look up [film]*. Netflix.

Meeker, J. W. (1974). *The comedy of survival: Studies in literary ecology*. Scribner.

Meier, M. R., & Berg, S. V. L. (2022). The activist potential of networked satire: Last week tonight with John Oliver and the struggle for net neutrality. *Western Journal of Communication, 88*(1), 3–32. https://doi.org/10.1080/10570314.2022.2130003

Meyers-Levy, J., & Malaviya, P. (1999). Consumers' processing of persuasive advertisements: An integrative framework of persuasion theories. *Journal of Marketing, 63*(4_suppl1), 45–60. https://doi.org/10.1177/00222429990634s106

Miéville, C. (2022). *Communism, the manifesto, and hate*. Salvage. https://salvage.zone/communism-the-manifesto-and-hate/

Milman, O. (2021). Climate denial is waning on the right. What's replacing it might be just as scary. *The Guardian*. https://www.theguardian.com/environment/2021/nov/21/climate-denial-far-right-immigration

Mish-Mash Ecologism. (2022). *New left review*. https://newleftreview.org/sidecar/posts/mish-mash-ecologism

Morris, C. W. (Ed.). (2003). *The social contract theorists: Critical essays on Hobbes, Locke, and Rousseau*. Rowman & Littlefield.

Morrison, S. (2017). Scott Morrison Brings in Lump of Coal Into House of Representatives. [Video]. YouTube. https://youtu.be/ea5bOaPkZpc

Morton, T. (2010). *The ecological thought*. Harvard University Press.

Moyer-Gusé, E., Robinson, M. J., & McKnight, J. (2018). The role of humor in messaging about the MMR vaccine. *Journal of Health Communication, 23*(6), 514–522. https://doi.org/10.1080/10810730.2018.1473533

Moyer-Gusé, E., Tchernev, J. M., & Walther-Martin, W. (2019). The persuasiveness of a humorous environmental narrative combined with an explicit persuasive appeal. *Science Communication, 41*(4), 422–441. https://doi.org/10.1177/1075547019862553

Nabi, R. L., Moyer-Gusé, E., & Byrne, S. (2007). All joking aside: A serious investigation into the persuasive effect of funny social issue messages. *Communication Monographs, 74*(1), 29–54. https://doi.org/10.1080/03637750701196896

Nagel, T. (1979). What is it like to be a bat? In T. Nagel (Ed.), *Mortal questions*. Cambridge University Press.

Nixon, R. (2011). *Slow violence and the environmentalism of the poor*. Harvard University Press. https://doi.org/10.2307/j.ctt2jbsgw

Nkrumah, K. (1965). *Neo-colonialism: The last stage of imperialism*. Thomas Nelson & Sons, Ltd. https://www.marxists.org/ebooks/nkrumah/nkrumah-neocolonialism.pdf

Norgaard, K. M. (2011). *Living in denial: Climate change, emotions, and everyday life*. MIT Press.

Norman, J., & Welchman, A. (2025). The politics of pessimism. In C. Marshall & S. Shapshay (Eds.), *The new Cambridge companion to schopenhauer*. Cambridge University Press.

Oliver, J. (2022). Carbon offsets: Last week tonight with John Oliver (HBO). YouTube. https://youtu.be/6p8zAbFKpW0

Oliver, J. (Host). (2025). *Last Week Tonight with John Oliver* (Season 12, Episode 23) [TV series episode]. HBO.

Panda, P. (2020). Indigenous humor in Thomas King's the back of the turtle: An ecocritical perspective. *Studies in American Humor, 6*(2), 323–340. https://doi.org/10.5325/studamerhumor.6.2.0323

Pannett, R. (2025, January 22). Colorado zoo sued over plans to bring elephants from Swaziland. *The Washington Post*. https://www.washingtonpost.com/climate-environment/2025/01/22/colorado-zoo-elephant-lawsuit/

Peiffer, K. S. (2000). *Coyote at large: Humor in American nature writing*. The University of Utah Press.

Peteinaki, M. (2024). *Reevaluating Sustainable Development Goals and European Climate Policies through the Doughnut Model*. [Master's dissertation, Hellenic Open University]. Apothesis Institutional Repository. https://apothesis.eap.gr/archive/item/212712?lang=en

Pérez, R. (2022). *The souls of white jokes: How racist humor fuels white supremacy*. NYU Press.

Petty, R. E., & Cacioppo, J. T. (1986). The elaboration likelihood model of persuasion. In *Advances in experimental social psychology* (pp. 123–205). Elsevier.

Pfattheicher, S., et al. (2016). The role of moral emotions in pro-environmental behavior: A literature review. *Environmental Science & Policy, 55*(1), 57–66.

Phiddian, R. (2020). *Satire and the public emotions*. Cambridge University Press.

Pini, M. (2022, July 21). Morrison courts Pentecostal world domination. *Independent Australia*. https://independentaustralia.net/politics/politics-display/morrison-courts-pentecostal-world-domination,16582

Plain, G. (2023). *Prosthetic agency: Literature, culture, and masculinity after World War II*. Cambridge University Press.

Rabaté, J.-M. (2016). Can you spare a laugh? Lacan, Freud, and Marx on the economy of jokes1. In P. Gherovici & M. Steinkoler (Eds.), *Lacan, psychoanalysis, and comedy* (pp. 82–103). Cambridge University Press.

Remnick, D. (2019). Alexandria Ocasio-Cortez is coming for your hamburgers! *The New Yorker*. https://www.newyorker.com/news/daily-comment/alexandria-ocasio-cortez-is-coming-for-your-hamburgers

Riofrancos, T. (2025). *Extraction: The frontiers of green capitalism*. W.W. Norton.

Robbins, P., & Moore, S. A. (2015, June 19). Love your symptoms: A sympathetic diagnosis of An ecomodernist manifesto. *Entitle Blog*. https://entitleblogdotorg3.wordpress.com/2015/06/19/love-your-symptoms-a-sympathetic-diagnosis-of-the-ecomodernist-manifesto/

Roberts, B. (2021). On 'Campaigning' for nuclear deterrence. In S. Stephan Frühling & A. O'Neil (Eds.), *Alliances, nuclear weapons and escalation: Managing deterrence in the 21st century* (1st ed., pp. 187–198). ANU Press. http://www.jstor.org/stable/j.ctv25m8dp0.21

Robinson, C. J. (2019). *Cedric J. Robinson: On racial capitalism, black internationalism, and cultures of resistance*. Pluto Press. https://doi.org/10.2307/j.ctvr0qs8p

Rogers, E. M. (1983). *Diffusion of innovations* (3rd ed.). Free Press of Glencoe.

Rogers, R. W. (1975). A protection motivation theory of fear appeals and attitude change. *The Journal of Psychology, 91*(1), 93–114. https://doi.org/10.1080/00223980.1975.9915803

Rorty, R. (1989). *Contingency, irony, and solidarity*. Cambridge University Press.
Rossing, J. P. (2013). Dick Gregory and activist style: Identifying attributes of humor necessary for activist advocacy. *Argumentation and Advocacy, 50*(2), 59–71. https://doi.org/10.1080/00028533.2013.11821810
Rossing, J. P. (2016). A sense of humor for civic life: Toward a strong defense of humor. *Studies in American Humor, 4*(1), 1–21. https://doi.org/10.5325/studamerhumor.2.1.0001
Rossing, J. P. (2019). An ethics of complicit criticism for postmodern satire. *Studies in American Humor, 5*(1), 13–30. https://doi.org/10.5325/studamerhumor.5.1.0013
Ruch, W., Heintz, S., Platt, T., Wagner, L., & Proyer, R. T. (2018). Broadening humor: Comic styles differentially tap into temperament, character, and ability. *Frontiers in Psychology, 9*, 6. https://doi.org/10.3389/fpsyg.2018.00006
Rupasinghe, R., Chomel, B. B., & Martínez-López, B. (2021). Climate change and zoonoses: A review of the current status, knowledge gaps, and future trends. *Acta Tropica, 226*, 106225. https://doi.org/10.1016/j.actatropica.2021.106225
Rushdie, S. (1988). *The satanic verses*. Vintage Books.
Sachs, A. (2023). *Stay cool: Why dark comedy matters in the fight against climate change*. NYU Press. http://www.jstor.org/stable/jj.8784649
Saito, K. (2024). *Slow down: The degrowth manifesto*. Astra Publishing House.
Salmose, N. (2018). The apocalyptic sublime: Anthropocene representation and environmental agency in Hollywood action-adventure Cli-Fi films. *The Journal of Popular Culture, 51*(6), 1415–1433. https://doi.org/10.1111/jpcu.12742
Schultz, P. W., & Kaiser, F. G. (2012). Promoting pro-environmental behavior. In S. D. Clayton (Ed.), *Oxford library of psychology. The Oxford handbook of environ-mental and conservation psychology* (pp. 556–580). Oxford University Press.
Schultz, P. W., Messina, A., Tronu, G., Limas, E. F., Gupta, R., & Estrada, M. (2016). Personalized normative feedback and the moderating role of personal norms: A field experiment to reduce residential water consumption. *Environment and Behavior, 48*(5), 686–710. https://doi.org/10.1177/0013916514553835
Schwartz, S. H. (1992). Universals in the content and structure of values: Theoretical advances and empirical tests in 20 countries. In M. Zanna (Ed.), *Advances in experimental social psychology* (25th ed., pp. 1–65). Academic Press. https://doi.org/10.1016/S0065-2601(08)60281-6
Schwartz, S. H. (1994). Are there universal aspects in the structure and contents of human values? *Journal of Social Issues, 50*(4), 19–45. https://doi.org/10.1111/j.1540-4560.1994.tb01196.x
Secreast, D. (2023). Carson's can of worms: Grotesque satire and abjection in *Rachel Carson's Silent Spring. ISLE: Interdisciplinary Studies in Literature and Environment, 30*(2), 450–469. https://doi.org/10.1093/isle/isab021

Selg, P., & Ventsel, A. (2020). The "relational turn" in the social sciences. In *Introducing relational political analysis. Palgrave studies in relational sociology*. Palgrave Macmillan. https://doi.org/10.1007/978-3-030-48780-5_2

Sels, A. J., Kuipers, G., & Aupers, S. (2025). *Don't look up*: Satirical cli-fi movies as a catalyst for online environmental debate. *European Journal of Cultural Studies*. https://doi.org/10.1177/13675494251318987

Seymour, N. (2018). *Bad environmentalism: Irony and irreverence in the ecological age*. University of Minnesota Press.

Shahab, M. H., Ghazali, E., & Mohtar, M. (2021). The role of elaboration likelihood model in consumer behaviour research and its extension to new technologies: A review and future research agenda. *International Journal of Consumer Studies, 45*(4), 664–689.

Shakespeare, W. (n.d.). *Hamlet*. MIT Shakespeare. Retrieved August 4, 2025, from https://shakespeare.mit.edu/hamlet/full.html

Shaw, T., & Youngblood, D. J. (2017). Cold war sport, film, and propaganda: A comparative analysis of the superpowers. *Journal of Cold War Studies, 19*(1), 160–192. https://www.jstor.org/stable/26925679

Shome, D., & Marx, S. (2009). *The psychology of climate change communication: A guide for scientists, journalists, educators, political aides, and the interested public*. Center for Research on Environmental Decisions.

Sienkiewicz, M., & Marx, N. (2021). Appropriating irony: Conservative comedy, Trump-era satire, and the politics of television humor. *Journal of Cinema and Media Studies, 60*(4), 85–108.

Sienkiewicz, M., & Marx, N. (2023). *That's not funny: How the right makes comedy work for them*. University of California Press.

Simon, W. (2003). Dr. Strangelove or: The apparatus of nuclear warfare. In R. Allen & M. Turvey (Eds.), *Camera obscura, camera lucida: Essays in honor of Annette Michelson* (pp. 215–230). Amsterdam University Press. https://doi.org/10.1515/9789048505067-013

Skiveren, N. (2024). Humor as hope? *Environmental Humanities, 16*(2), 441–459. https://doi.org/10.1215/22011919-11150075

Skurka, C., Niederdeppe, J., & Nabi, R. (2019). Kimmel in climate: Disentangling the emotional ingredients of satirical monologue. *Science Communication, 41*(4), 394–421. https://doi.org/10.1177/1075547019853837

Skurka, C., Niederdeppe, J., Romero-Canyas, R., & Acup, D. (2018). Pathways of influence in emotional appeals: Benefits and tradeoffs of using fear or humor to promote climate change-related intentions and risk perceptions. *Journal of Communication, 68*(1), 169–193. https://doi.org/10.1093/joc/jqx008

Skurka, C., Romero-Canyas, R., Joo, H. H., Acup, D., & Niederdeppe, J. (2022). Emotional appeals, climate change, and young adults: A direct replication of Skurka et al. (2018). *Human Communication Research, 48*(1), 147–156. https://doi.org/10.1093/hcr/hqab013

Solnit, R. (2016). *Hope in the dark: Untold histories, wild possibilities* (3rd ed.). Haymarket Books.

Sørensen, M. (2016). *Humor in political activism creative nonviolent resistance.* Palgrave Macmillan.

Sparrow, J. (2019, March 10). Against the new vitalism. *New Socialist.* https://newsocialist.org.uk/against-the-new-vitalism/

Sperb, J. (2006). *The Kubrick facade: Faces and voices in the films of Stanley Kubrick.*

Sperrin, D. (2025). *State of ridicule: A history of satire in English literature.* Princeton University Press. https://doi.org/10.2307/jj.22679818

Srnicek, N. (2017). *Platform capitalism.* Polity Press.

Standing, G. (2011). *The precariat: The new dangerous class.* Bloomsbury Academic.

Steward, H. (2018). Akrasia. In *Routledge eBooks.* https://doi.org/10.4324/9780415249126-v003-1

Steg, L., Bolderdijk, J. W., Keizer, K., & Perlaviciute, G. (2014). An integrated frame-work for encouraging pro-environmental behavior: The role of values, situational factors and goals. *Journal of Environmental Psychology, 38*, 104–115. https://doi.org/10.1016/j.jenvp.2014.01.002

Steg, L., & Nordlund, A. (2012). Models to explain environmental behavior. In L. Steg, A. E. van den Berg, & J. I. M. de Groot (Eds.), *Environmental psychology: An introduction* (pp. 185–195). BPS Blackwell.

Steg, L., & Nordlund, A. (2018). Theories to explain environmental behavior. In L. Steg & J. de Groot (Eds.), *Environmental psychology: An introduction* (2nd ed., pp. 217–227). Wiley. https://doi.org/10.1002/9781119241072.ch22

Steg, L., Perlaviciute, G., van der Werff, E., & Lurvink, J. (2014). The significance of hedonic values for environmentally relevant attitudes, preferences, and actions. *Environment and Behavior, 46*(2), 163–192. https://doi.org/10.1177/0013916512454730

Stillman, G. B. (2008). Two of the MADdest scientists: Where Strangelove Meets Dr. No; or, unexpected roots for Kubrick's Cold war classic. *Film History, 20*(4), 487–500. http://www.jstor.org/stable/27670748

Strick, M. (2021). How humor can promote central-route persuasion: The role of ambivalence. In M. Strick & T. E. Ford (Eds.), *The social psychology of humor* (pp. 3–19). Routledge.

Strick, M., & Ford, T. E. (Eds.). (2021). *The social psychology of humor* (1st ed.). Routledge. https://doi.org/10.4324/9781003042440

Sultana, F. (Ed.). (2024). *Confronting climate coloniality: Decolonizing pathways for climate justice* (1st ed.). Routledge. https://doi.org/10.4324/9781003465973

Swift, J. (2005). *Gulliver's travels.* Edited with an introduction by Claude Rawson and notes by Ian Higgins. Oxford and New York: Oxford University Press.

Swim, J. K., Markowitz, E. M., & Bloodhart, B. (2012). Psychology and climate change: Beliefs, impacts, and human contributions. In S. D. Clayton (Ed.), *The Oxford handbook of environmental and conservation psychology.* Oxford University Press. https://doi.org/10.1093/oxfordhb/9780199733026.013.0033

Takach, G. (2022). Eco-comedy. *Environmental Humanities, 14*(2), 371–374. https://doi.org/10.1215/22011919-9712445

Test, G. A. (1991). *Satire: Spirit and art.* University Press of Florida.

The Breakthrough Institute. (2015, July 14). *What is modern in ecomodernism? Nature, technology, and politics in the Anthropocene.* https://thebreakthrough.org/articles/what-is-modern-in-ecomodernism

The Late Show with Stephen Colbert. (2025, September 26). Meanwhile... Central Perk NYC | Disney Scuba Robbery | Crackless Tomatoes | Coffee makes you happy [Video]. YouTube. https://www.youtube.com/watch?v=NDkjBGiUZig

The Onion. (2025a). *Court rules elephants can't sue to leave zoo. The Onion.* https://theonion.com/court-rules-elephants-cant-sue-to-leave-zoo/

The Onion. (2025b). *Orcas begin bringing humans gifts of food in hopeful first step toward peace.* https://theonion.com/orcas-bringing-humans-gifts-of-food/

Thompson, A. D. (2025, June 30). Daily cartoon: Monday, June 30th [Cartoon]. *The New Yorker.* https://www.newyorker.com/cartoons/daily-cartoon/monday-june-30th-pre-apocalyptic-fiction

Tilley, L. (n.d.). *Indonesian nickel sovereignty and global raced finance: The meanings of element 28 in an age of transition.*

Tonn, M. B. (2017). Making and unmaking political mischief: Trickster influences in the rhetorical humor of the 1960s. In R. J. Jensen (Ed.), *Social controversy and public address in the 1960s and early 1970s: A rhetorical history of the United States* (Vol. 9, pp. 221–250). Michigan State University Press. https://doi.org/10.14321/j.ctt1vjqqkq.11

Transport & Environment. (2025, October). Smoke screen: the growing PHEV emissions scandal — Long-range PHEVs and EREVs are a diversion on the road to zero emissions. Authors: Sofía Navas Gohlke, Yoann Gimbert. Expert group: Lucien Mathieu, Richard Riley. Editor responsible: William Todts. Brussels: European Federation for Transport and Environment AISBL. Retrieved from https://www.transportenvironment.org/uploads/files/2025_10_PHEV_smoke_screen_report.pdf

Traub, C. (2018). From the grotesque to nuclear-age precedents: The modes and meanings of cli-fi humor. *Studies in the Novel, 50*(1), 86–107. https://www.jstor.org/stable/48559258

United Nations Framework Convention on Climate Change. (n.d.). *Kyoto protocol.* https://unfccc.int/kyoto_protocol

United Nations Framework Convention on Climate Change (UNFCCC). (n.d.). *[Official website].* https://unfccc.int/

Van der Lugt, M. (2025). *Hopeful pessimism.* Princeton University Press.

Vanderheiden, E., & Mayer, C. H. (2025). Investigating shame as a resource in the anthropocene: An emotional psychology perspective. In C. H. Mayer & E. Vanderheiden (Eds.), *International handbook of emotions.* Springer. https://doi.org/10.1007/978-3-031-86449-0_3

Varoufakis, Y. (2023). *Technofeudalism: What killed capitalism.* Penguin.

Wagner, P. (2023). Embracing the environmental grotesque and transforming the climate crisis. *ISLE: Interdisciplinary Studies in Literature and Environment, 30*(4), 911–930. https://doi.org/10.1093/isle/isab086

Walker, K. (2021). As climate displacement increases, migration myths fuel fears. *Equal Times*. https://www.equaltimes.org/as-climate-displacement-increases

Warren, K. J. (2000). *Ecofeminist philosophy: A Western perspective on what it is and why it matters*. Rowman & Littlefield.

West, S., Haider, L. J., Stålhammar, S., & Woroniecki, S. (2020). A relational turn for sustainability science? Relational thinking, leverage points and transformations. *Ecosystems and People, 16*(1), 304–325. https://doi.org/10.1080/26395916.2020.1814417

White, H. (1973). *Metahistory: The historical imagination in nineteenth-century Europe*. Johns Hopkins University Press.

Wiek von Mossner, A. (2020). Affect, emotion, and ecocriticism. *Ecozon@ European Journal of Literature Culture and Environment, 11*(2), 128–136.

Wilderson, F. B. (2020). *Afropessimism*. Liveright.

Willett, C., & Willett, J. A. (2019). *Uproarious: How feminists and other subversive comics speak truth*. University of Minnesota Press.

Winterscheidt, J. (Host). (2023). *The world's most dangerous show with Joko Winterscheidt* [TV series]. Amazon Prime Video.

World People's Conference on Climate Change and the Rights of Mother Earth. (2010). *The People's Agreement of Cochabamba. Cochabamba, Bolivia*. https://pwccc.wordpress.com/2010/04/24/peoples-agreement

Wyer, R. S., & Collins, J. E. (1992). A theory of humor elicitation. *Psychological Review, 99*(4), 663–688. https://doi.org/10.1037/0033-295X.99.4.663

Yang, Q., & Lee, Y.-C. (2023). The effect of live streaming commerce quality on customers' purchase intention: Extending the elaboration likelihood model with herd behaviour. *Behaviour & Information Technology, 43*(5), 907–928.

Yeo, S. K., Becker, A. B., Cacciatore, M. A., Anderson, A. A., & Patel, K. (2022). Humor can increase perceived communicator effectiveness regardless of race, gender, and expertise—If you are funny enough. *Science Communication, 44*(5), 593–620. https://doi.org/10.1177/10755470221132278

Young, D. G. (2008). The privileged role of the late-night joke: Exploring humor's role in disrupting argument scrutiny. *Media Psychology, 11*(1), 119–142. https://doi.org/10.1080/15213260701837073

Young, D. G. (2020). *Irony and outrage: The polarized landscape of rage, fear, and laughter in the United Sates*. Oxford University Press. https://doi.org/10.1093/poq/nfaa020

Young, J. R. (2016). *A comedian and an academic walk into a podcast*. Chronicle of Higher Education. www.chronicle.com/article/A-Comedianan-Academic/237715

Yousfi, F. (n.d.). *Crisis in the fields: Neoliberal agricultural policies and rural discontent in Morocco*. Ghent University.

Zekavat, M. (2014). A discursive model of satire. *JESELL: Jena Electronic Stud-ies in English Language and Literature*. http://zs.thulb.uni-jena.de/receive/jportal_jparticle_00324293

Zekavat, M. (2017). *Satire, humor and the construction of identities*. John Benjamins.

Zekavat, M. (2019a). Reflexive humour and satire: A critical review. *European Journal of Humour Research, 7*(4), 125–136. https://doi.org/10.7592/EJHR 2019.7.4.zekavat

Zekavat, M. (2019b). Satire, humor and ecological thought. *Neohelicon, 46*(3), 369–386.

Zekavat, M. (2023a). The ambivalent affordances of humour in capitalist organisations. *European Journal of Humour Research, 11*(1), 184–200. https://doi.org/10.7592/EJHR2023.11.1.728

Zekavat, M. (2023b). The ambivalent affordances of humour in capitalist organizations. *The European Journal of Humour Research, 11*(1), 184–200. https://doi.org/10.7592/ejhr.2023.11.1.728

Zekavat, M. (2025a). Communicating multispecies planetary health through multimodal humour and satire. *Medical Humanities*, medhum-013359. https://doi.org/10.1136/medhum-2025-013359

Zekavat, M. (2025b). When aporetic satire backfires: Chappelle's transgender jokes gone awry. *Symplokē, 34*, 1–2.

Zekavat, M. (2025c). The paradox of political satire: Navigating critique in culture industry and neoliberal media. *Comedy Studies*. https://doi.org/10.1080/2040610X.2025.2509429

Zekavat, M. (2026a). When aporetic satire backfires: Chappelle's transgender jokes gone awry. *Symplokē, 34*, 1–2.

Zekavat, M. (Forthcoming-b). Planetary health and emancipatory worlding. *Global Health Promotion*. http://dx.doi.org/10.1177/17579759251378432

Zekavat, M. (Forthcoming-c). Cartooning in the anthropocene: Exposing the environmental impacts of capitalism.

Zekavat, M. (Forthcoming-d). Planetary health and emancipatory worlding.

Zekavat, M. (Forthcoming-e). Reimagining resistance: Resource scarcity, gender, and colonial invasion in Simin Daneshvar's Savushun.

Zekavat, M. (Forthcoming-f). Rumours and realities: The role of satire in navigating the blah blah blah of environmental politics.

Zekavat, M., & Scheel, T. (2023). *Satire, humor, and environmental crises* (1st ed.). Routledge.

ŽiŽek, S. (2006). *The pervert's guide to cinema: 1, 2, 3* [Film]. Directed by S. Fiennes. Lone Star, Mischief Films, Amoeba Film.

# Index[1]



[1] Note: Page numbers followed by 'n' refer to notes.


M. Zekavat, *Leveraging Satire for Environmental Advocacy*, https://doi.org/10.1007/978-3-032-04838-7

**D**

FSC
www.fsc.org
MIX
Papier | Fördert
gute Waldnutzung
FSC® C083411

Kolibri 360 GmbH
Ferdinand-Jühlke-Straße 7
99095 Erfurt, Deutschland
produktsicherheit@kolibri360.de

FSC
www.fsc.org
MIX
Papier | Fördert gute Waldnutzung
FSC® C083411

Kolibri 360 GmbH
Ferdinand-Jühlke-Straße 7
99095 Erfurt, Deutschland
produktsicherheit@kolibri360.de